HOLT
Biology

Lab Manual for Skills Practice Labs and Inquiry Labs, Level B

HOLT, RINEHART AND WINSTON
A Harcourt Education Company
Orlando • Austin • New York • San Diego • London

Copyright © by Holt, Rinehart and Winston

All rights reserved. No part of this publication may be reproduced or transmitted in any form or by any means, electronic or mechanical, including photocopy, recording, or any information storage and retrieval system, without permission in writing from the publisher.

Teachers using HOLT BIOLOGY may photocopy blackline masters in complete pages in sufficient quantities for classroom use only and not for resale.

CBL is a trademark of Texas Instruments Incorporated.

HOLT and the **"Owl Design"** are trademarks licensed to Holt, Rinehart and Winston, registered in the United States of America and/or other jurisdictions.

Printed in the United States of America.

If you have received these materials as examination copies free of charge, Holt, Rinehart and Winston retains title to the materials and they may not be resold. Resale of examination copies is strictly prohibited.

Possession of this publication in print format does not entitle users to convert this publication, or any portion of it, into electronic format.

ISBN-13: 978-0-03-093217-5
ISBN-10: 0-03-093217-3

3 4 5 6 7 862 10 09 08

Contents

Lab Safety 1

Safety Symbols 4

Biology and You
Skills Practice Lab: SI Units 7

Applications of Biology
Skills Practice Lab: Microbe Growth 12

Chemistry of Life
Inquiry Lab: Enzymes in Detergents 16

Ecosystems
Inquiry Lab: Ecosystem Change 20

Populations and Communities
Skills Practice Lab: Yeast Population Growth 27

The Environment
Inquiry Lab: Effects of Acid Rain on Seeds 31

Cell Structure
Skills Practice Lab: Plant Cell Observation 36

Cells and Their Environment
Inquiry Lab: Cell Size and Diffusion 38

Photosynthesis and Cellular Respiration
Skills Practice Lab: Cellular Respiration 43

Cell Growth and Division
Skills Practice Lab: Mitosis in Plant Cells 46

Meiosis and Sexual Reproduction
Skills Practice Lab: Meiosis Model 50

Mendel and Heredity
Inquiry Lab: Plant Genetics 53

DNA, RNA, and Proteins
Skills Practice Lab: DNA Extraction from Wheat Germ 55

Genes in Action
Skills Practice Lab: Protein Detection 57

Gene Technologies and Human Applications
Skills Practice Lab: DNA Fingerprint Analysis 61

Evolutionary Theory
Skills Practice Lab: Natural Selection Simulation 66

Population Genetics and Speciation
Skills Practice Lab: Genetic Drift 72

Classification
Skills Practice Lab: Dichotomous Keys 75

Contents

History of Life on Earth
Skills Practice Lab: Model of Rock Strata 79

Bacteria and Viruses
Skills Practice Lab: Bacterial Staining 81

Protists
Inquiry Lab: Protistan Responses to Light 86

Fungi
Inquiry Lab: Yeast and Fermentation 92

Plant Diversity and Life Cycles
Skills Practice Lab: Plant Diversity 95

Seed Plant Structure and Growth
Skills Practice Lab: Monocot and Dicot Seeds 99

Plant Processes
Skills Practice Lab: Cultivation Techniques 103

Introduction to Animals
Skills Practice Lab: Embryonic Development 107

Simple Invertebrates
Skills Practice Lab: Hydra Behavior 112

Mollusks and Annelids
Skills Practice Lab: Clam Characteristics 114

Arthropods and Echinoderms
Skills Practice Lab: Butterfly Metamorphosis 118

Fishes and Amphibians
Skills Practice Lab: Live Frog Observation 123

Reptiles and Birds
Skills Practice Lab: Bird Digestion 127

Mammals
Skills Practice Lab: Mammalian Characteristics 131

Animal Behavior
Inquiry Lab: Territorial Behavior 136

Skeletal, Muscular, and Integumentary Systems
Inquiry Lab: Analysis of Muscle Fatigue 139

Circulatory and Respiratory Systems
Skills Practice Lab: Lung Capacity 142

Digestive and Excretory Systems
Inquiry Lab: Lactose Digestion 146

The Body's Defenses
Skills Practice Lab: Disease Transmission Model 150

Contents

Nervous System
Inquiry Lab: Reaction Times 154

Endocrine System
Inquiry Lab: Epinephrine and Heart Rate 159

Reproduction and Development
Skills Practice Lab: Sonography 163

Forensic Science
Skills Practice Lab: The Counterfeit Drugs 166

Name _____ Class _____ Date _____

Lab Safety

In the laboratory or in the field, you can engage in hands-on explorations, test your scientific hypotheses, and build practical lab skills. However, while you are working, it is your responsibility to protect yourself and your classmates by conducting yourself in a safe manner. You will avoid accidents by following directions, handling materials carefully, and taking your work seriously. Read the following safety guidelines before working in the lab or field. Make sure that you understand all safety guidelines before entering the lab or field.

Before You Begin

- **Read the entire activity before entering the lab.** Be familiar with the instructions before beginning an activity. Do not start an activity until you have asked your teacher to explain any parts of the activity that you do not understand.

- **Student-designed procedures or inquiry activities must be approved by your teacher before you attempt the procedures or activities.**

- **Wear the right clothing for lab work.** Before beginning work, tie back long hair, roll up loose sleeves, and put on any required personal protective equipment as directed by your teacher. Remove all jewelry, and confine all clothing that could knock things over, catch on fire, contact electrical connections, or absorb chemical solutions. Wear pants rather than shorts or skirts. Protect your feet from chemical spills and falling objects. Do not wear open-toed shoes, sandals, or canvas shoes in the lab. In addition, chemical fumes may react with and ruin some jewelry, such as pearl jewelry. Do not apply cosmetics in the lab. Some hair care products and nail polish are highly flammable.

- **Do not wear contact lenses in the lab.** Even though you will be wearing safety goggles, chemicals could get between contact lenses and your eyes and could cause irreparable eye damage. If your doctor requires that you wear contact lenses instead of glasses, then you should wear eye-cup safety goggles—similar to goggles worn for underwater swimming—in the lab. Ask your doctor or your teacher how to use eye-cup safety goggles to protect your eyes.

- **Know the location of all safety and emergency equipment used in the lab.** Know proper fire-drill procedures and the location of all fire exits. Ask your teacher where the nearest eyewash stations, safety blankets, safety shower, fire extinguisher, first-aid kit, and chemical spill kit are located. Be sure that you know how to operate the equipment safely.

While You Are Working

- **Always wear a lab apron and safety goggles.** Wear these items while in the lab, even if you are not working on an activity. Labs contain chemicals that can damage your clothing, skin, and eyes. Aprons and goggles also protect against many physical hazards. If your safety goggles cloud up or are uncomfortable, ask your teacher for help. Lengthening the strap slightly, washing the goggles with soap and warm water, or using an anti-fog spray may help the problem.

Name _____ Class _____ Date _____
Lab Safety *continued*

- **NEVER work alone in the lab.** Work in the lab only when supervised by your teacher.
- **NEVER leave equipment unattended while it is in operation.**
- **Perform only activities specifically assigned by your teacher.** Do not attempt any procedure without your teacher's direction. Use only materials and equipment listed in the activity or authorized by your teacher. Steps in a procedure should be performed only as described in the activity or as approved by your teacher.
- **Keep your work area neat and uncluttered.** Have only books and other materials that are needed to conduct the activity in the lab. Keep backpacks, purses, and other items in your desk, your locker, or other designated storage areas.
- **Always heed safety symbols and cautions listed in activities, listed on handouts, posted in the room, provided on equipment or chemical labels (whether provided by the manufacturer or added later), and given verbally by your teacher.** Be aware of the potential hazards of the required materials and procedures, and follow all precautions indicated.
- **Be alert, and walk with care in the lab.** Be aware of others near you and your equipment, and be aware of what they are doing.
- **Do not take food, drinks, chewing gum, or tobacco products into the lab.** Do not store or eat food in the lab. Either finish these items or discard them before coming into the lab or beginning work in the field.
- **NEVER taste chemicals or allow them to contact your skin.** Keep your hands away from your face and mouth, even if you are wearing gloves. Only smell vapors as instructed by your teacher and only in the manner indicated.
- **Exercise caution when working with electrical equipment.** Do not use electrical equipment with frayed or twisted wires. Check that insulation on wiring is intact. Be sure that your hands are dry before using electrical equipment. Do not let electrical cords dangle from work stations. Dangling cords can catch on apparatus on tables, can cause you to trip, and can cause an electric shock. The area under and around electrical equipment should be dry; cords should not lie in puddles of spilled liquid, under sink spigots, or in sinks themselves.
- **Use extreme caution when working with hot plates and other heating devices.** Keep your head, hands, hair, and clothing away from the flame or heating area. Never leave a heating device unattended when it is in use. Metal, ceramic, and glass items do not necessarily look hot when they are hot. Allow all items to cool before storing them.
- **Guard against complacency.** Remember that it is human nature to become careless when doing routine things. As you become familiar with apparatus and procedures, remain alert and pay attention.
- **Do not fool around in the lab.** Take your lab work seriously, and behave appropriately in the lab. Lab equipment and apparatus are not toys; never use lab time or equipment for anything other than the intended purpose. Be considerate and be aware of the safety of your classmates as well as your safety at all times.

Original content Copyright © by Holt, Rinehart and Winston. Additions and changes to the original content are the responsibility of the instructor.

Name _____ Class _____ Date _____
Lab Safety *continued*

Emergency Procedures

- **Follow standard fire-safety procedures.** If your clothing catches on fire, do not run; WALK to the safety shower, stand under it, and turn it on. While doing so, call to your teacher.

- **Report any accident, incident, or hazard—no matter how trivial—to your teacher immediately.** Any incident involving bleeding, burns, fainting, nausea, dizziness, chemical exposure, or ingestion should also be reported immediately to the school nurse or to a physician. If you have a close call, tell your teacher so that you and your teacher can find a way to prevent it from happening again.

- **Report all spills to your teacher immediately.** Call your teacher rather than trying to clean a spill yourself. Your teacher will tell you whether it is safe for you to clean up the spill; if it is not safe, your teacher will know how to clean up the spill.

- **If you spill a chemical on your skin, wash the chemical off in the sink and call your teacher.** If you spill a solid chemical onto your clothing, using an appropriate container, brush it off carefully without scattering it onto somebody else and call your teacher. If you spill corrosive substances on your skin or clothing, use the safety shower or a faucet to rinse. Remove affected clothing while you are under the shower, and call to your teacher. (It may be temporarily embarrassing to remove clothing in front of your classmates, but failure to thoroughly rinse a chemical off your skin could result in permanent damage.)

- **If you get a chemical in your eyes, walk immediately to the eyewash station, turn it on, and lower your head so that your eyes are in the running water.** Hold your eyelids open with your thumbs and fingers, and roll your eyeballs around. You have to flush your eyes continuously for at least 15 minutes. Call your teacher while you are flushing your eyes.

When You Are Finished

- **Clean your work area at the conclusion of each lab period as directed by your teacher.** Broken glass, chemicals, and other waste products should be disposed of in separate, special containers. Dispose of waste materials as directed by your teacher. Put away all material and equipment according to your teacher's instructions. Report any damaged or missing equipment or materials to your teacher.

- **Even if you wore gloves, wash your hands with soap and hot water after each lab period.** To avoid contamination, wash your hands at the conclusion of each lab period and before you leave the lab.

Name _____ Class _____ Date _____

Safety Symbols

Before you begin working on an activity, familiarize yourself with the following safety symbols, which are used throughout your textbook, and the guidelines that you should follow when you see these symbols.

 EYE PROTECTION

- **Wear approved safety goggles as directed.** Safety goggles should be worn in the lab at all times, especially when you are working with a chemical or solution, a heat source, or a mechanical device.
- **If chemicals get into your eyes, flush your eyes immediately.** Go to an eyewash station immediately, and flush your eyes (including under the eyelids) with running water for at least 15 minutes. Use your thumb and fingers to hold your eyelids open, and roll your eyeballs around. While doing so, call your teacher or ask another student to notify your teacher.
- **Do not wear contact lenses in the lab.** Chemicals can be drawn up under a contact lens and into the eye. If you must wear contacts prescribed by a physician, tell your teacher. In this case, you must also wear approved eye-cup safety goggles.
- **Do not look directly at the sun or any intense light source or laser.** Do not reflect direct sunlight to illuminate a microscope. Such action concentrates light rays to an intensity that can severely burn your retinas and cause blindness.

 CLOTHING PROTECTION

- **Wear an apron or lab coat at all times in the lab to prevent chemicals or chemical solutions from contacting skin or clothes.**
- **Tie back long hair, secure loose clothing, and remove loose jewelry so that they do not knock over equipment or come into contact with hazardous materials or electrical connections.**
- **Do not wear open-toed shoes, sandals, or canvas shoes in the lab.** Splashed chemicals directly contact skin or quickly soak through canvas. Hard shoes will not allow chemicals to soak through as quickly, and they provide more protection against dropped or falling objects.

 HAND SAFETY

- **Wear appropriate protective gloves when working with a heat source, chemicals, solutions, or wild or unknown plants.** Your teacher will provide the type of gloves necessary for a given activity.

 SHARP-OBJECT SAFETY

- **Use extreme care when handling all sharp and pointed instruments, such as scalpels, sharp probes, and knives.**
- **Do not cut an object while holding the object in your hand.** Cut objects on a suitable work surface. Always cut in a direction away from your body.
- **Do not use double-edged razor blades in the lab.**

Original content Copyright © by Holt, Rinehart and Winston. Additions and changes to the original content are the responsibility of the instructor.

Holt Biology — Safety Symbols

Safety Symbols *continued*

GLASSWARE SAFETY

- **Inspect glassware before use; do not use chipped or cracked glassware.** Use heat-resistant glassware for heating materials or storing hot liquids, and use appropriate tongs or a heat-resistant mitt to handle this equipment.

- **Notify your teacher immediately if a piece of glassware or a light bulb breaks.** Do not attempt to clean up broken glass or remove broken bulbs unless your teacher directs you to do so.

CHEMICAL SAFETY

- **Always wear safety goggles, gloves, and a lab apron or coat to protect your eyes and skin when you are working with any chemical or chemical solution. Do not taste, touch, or smell any chemicals or bring them close to your eyes unless specifically instructed to do so by your teacher.**

- **Know where the emergency lab shower and eyewash stations are and how to use them.** If you get a chemical on your skin or clothing, wash it off while calling to your teacher.

- **Handle chemicals or chemical solutions with care.** Check the labels on bottles, and observe safety procedures. Label beakers, flasks, test tubes, or other temporary storage vessels containing chemicals.

- **For all chemicals, take only what you need.** Do not return unused chemicals to their original containers.

- **NEVER take any chemicals out of the lab.**

- **Do not mix any chemicals unless specifically instructed to do so by your teacher.** Check the labels to make sure that you picked up the correct chemicals before you mix them. Otherwise harmless chemicals can be poisonous or explosive if combined.

- **Report all spills to your teacher immediately.** Clean up spills promptly as directed by your teacher.

ELECTRICAL SAFETY

- **Do not use equipment with frayed electrical cords or loose plugs.** Do not attempt to remove a plug tine if it breaks off in the socket. Notify your teacher, and stay away from the outlet.

- **Fasten electrical cords to work surfaces by using tape.** Doing so will prevent tripping and will ensure that equipment will not fall or be pulled off the table.

- **Do not use electrical equipment near water or when your clothing or hands are wet. Hold the plug housing when you plug in or unplug equipment.** Do not touch the metal prongs of the plug, and do not pull on the cord.

HEATING SAFETY

- **Avoid using open flames.** If possible, work only with hot plates that have an on/off switch and an indicator light. Turn off hot plates and open flames when they are not in use.

- **Never leave a hot plate unattended while it is turned on or while it is cooling off.**

- **Know the location of lab fire extinguishers and fire-safety blankets.**

- **Wear chemical splash goggles when working with liquids hotter than 60°C.**

Safety Symbols *continued*

- **Use tongs or appropriate insulated holders when handling heated objects.** Heated objects often do not appear to be hot. Do not pick up an object with your hand if the object could be warm.

- **Keep flammable substances away from heat, flames, and other ignition sources.** • **Allow all equipment to cool before storing it.**

ANIMAL CARE AND SAFETY

- **Handle animals only as directed by your teacher.** Mishandling or abusing animals will not be tolerated.

- **Always get your teacher's permission before bringing any animal (including pets) into the school building.**

- **Do not approach or touch any wild animals.** When working outdoors, be aware of poisonous or dangerous animals in the area.

- **Dispose of specimens only as instructed by your teacher.**

PLANT SAFETY

- **Do not ingest any plant part used in the laboratory (especially commercially sold seeds).** Do not touch any sap or plant juice directly. Always wear gloves.

- **Wear disposable polyethylene gloves when handling any wild plant.**

- **Wash hands thoroughly after handling any plant or plant part (particularly seeds).** Avoid touching your face and eyes.

- **Do not pick wild flowers or other plants unless instructed to do so by your teacher.**

HYGIENIC CARE

- **Keep your hands away from your face, hair, and mouth while you are working on any activity.**

- **Wash your hands thoroughly before you leave the lab or when you finish any activity.**

- **Remove contaminated clothing immediately.** If you spill corrosive substances on your skin or clothing, use the safety shower or a faucet to rinse. Remove affected clothing while you are under the shower, and call to your teacher. (It may be temporarily embarrassing to remove clothing in front of your classmates, but failure to thoroughly rinse a chemical off your skin could result in permanent damage.)

- **Use the proper technique demonstrated by your teacher when you are handling bacteria or other microorganisms.** Treat all microorganisms as if they are pathogens. Do not open Petri dishes to observe or count bacterial colonies.

PROPER WASTE DISPOSAL

- **Clean and sanitize all work surfaces and personal protective equipment after each lab period as directed by your teacher.**

- **Dispose of contaminated materials (biological or chemical) in special containers only as directed by your teacher.** Never put these materials into a regular waste container or down the drain.

- **Dispose of sharp objects (such as broken glass) in the appropriate sharps or broken glass container as directed by your teacher.**

Name _____ Class _____ Date _____

Skills Practice

DATASHEET B FOR IN-TEXT LAB

SI Units

OBJECTIVES

- **Express** measurements in SI units.
- **Read** a thermometer.
- **Measure** liquid volume by using a graduated cylinder.
- **Measure** mass by using a balance.
- **Determine** the density (mass-to-volume ratio) of two liquids.

MATERIALS

- graduated cylinder, 100 mL
- cups, plastic, (2)
- thermometers, Celsius, alcohol-filled (2)
- ring stand or lamp support
- stopwatch or clock
- corn oil, 25 mL
- cup, clear plastic
- sand, light-colored, 75 mL
- sand, dark-colored, 75 mL
- gloves, heat-resistant
- light source
- balance
- water, 25 mL
- graph paper

Procedure

MEASURE SAND TEMPERATURE

1. Use the data table on the following page to record your results.
2. Put on safety goggles, gloves, and a lab apron. Using a graduated cylinder, measure 75 mL of light-colored sand, and pour it into one of the small plastic cups. Repeat this procedure with the dark-colored sand and another plastic cup.
3. Level the sand by placing the cup on your desk and sliding the cup back and forth. Insert one thermometer into each cup. The zero line on the thermometer should be level with the sand, as shown in the picture on page 20 of your textbook. Relevel the sand if necessary.
4. Using a ring stand or lamp support, position the lamp approximately 9 cm from the top of the sand, as shown in the picture on page 20. Make sure that the lamp is evenly positioned between the two cups.

Name _____ Class _____ Date _____

SI Units *continued*

5. Before turning on the lamp, record in your data table the initial temperature of each cup of sand.

Sand Temperature

Time (min)	Temperature (degrees C)	
	Dark-colored sand	Light-colored sand
Start		
1		
2		
3		
4		
5		
6		
7		
8		
9		
10		

6. **CAUTION: Wear heat-resistant gloves when handling the lamp. The lamp will become very hot and may burn you.** Note the time or start the stopwatch when you turn on the lamp. The lamp will become hot and warm the sand. Check the temperature of the sand in each container at one-minute intervals for 10 minutes. In your data table, record the temperature of the sand after each minute.

COMPARE THE DENSITY OF OIL AND WATER

7. Use the data table on the following page to record the results from this section.

8. Label one clean plastic cup "Oil," and label another "Water." Using a balance, measure the mass of each plastic cup, and record the value in your data table.

Name _____ Class _____ Date _____
SI Units *continued*

Density of Two Liquids

a. Mass of empty oil cup		g
b. Mass of empty water cup		g
c. Mass of cup and oil		g
d. Mass of cup and water		g
e. Volume of oil		25 mL
f. Volume of water		25 mL
Calculating Actual Mass		
Oil Item c – Item a =		g
Water Item d – Item b =		g
g. Density of oil		g/mL
h. Density of water		g/mL

9. Put on an apron. Using a clean graduated cylinder, measure 25 mL of corn oil, and pour it into the plastic cup labeled "Oil." Using a balance, measure the mass of the plastic cup containing the corn oil, and record the mass in your data table.

10. Repeat step 9 with water and the plastic cup labeled "Water."

11. To find the mass of the oil, subtract the mass of the empty cup from the mass of the cup and the oil together.

12. To find the density of the oil, divide the mass of the oil by the volume of the oil, as shown in the operation below.

    $$\text{Density of oil} = \frac{\text{mass of oil}}{\text{volume of oil}} = \underline{\qquad} \text{ g/mL}$$

13. Repeat steps 11 and 12 to find the mass and density of water.

14. Combine the oil and water in the clear cup, and record your observations.

_____

_____

15. Clean up your materials according to your teacher's instructions. Wash your hands before leaving the lab.

Name _____ Class _____ Date _____

SI Units *continued*

Analyze and Conclude

1. **Graphing Data** Use graph paper or a graphing calculator to graph the data that you collected in the first part of the lab. Plot time on the *x*-axis and temperature on the *y*-axis.

2. **Scientific Methods Interpreting Data** Based on your graph, what is the relationship between color and heat absorption?

   _____

   _____

3. **Inferring Conclusions** How might the color of the clothes that you wear affect how warm you are on a sunny day?

   _____

   _____

4. **Scientific Methods Making Systematic Observations** In the second part of the lab, you combined the oil and water. Relate your observation to the densities that you calculated.

   _____

   _____

5. **Scientific Methods Using Evidence to Make Explanations** What could you infer about the value for the density of ice if you observe it floating in water?

   _____

   _____

Name _____ Class _____ Date _____
SI Units *continued*

Extensions

6. **Understanding Relationships** How would your calculated density values be affected if you misread the volume measurement on the graduated cylinder?

   _____

   _____

   _____

7. **Experimental Design** Pumice is a volcanic rock that has a density less than 1.00 g/cm³. How would you prove this density if you did not have a balance to weigh the pumice? (Hint: The density of water is 1.00 g/cm³.)

   _____

   _____

   _____

   _____

Name_____ Class_____ Date_____

Skills Practice

DATASHEET B FOR IN-TEXT LAB

Microbe Growth

OBJECTIVES

- **Observe** the effect of sterile and nonsterile conditions on the growth of microbes.
- **Make** daily observations, and organize data.
- **Hypothesize** about the conditions under which microbes grow.

MATERIALS

- pencil, wax
- broth, nutrient
- water, sterilized
- gloves, heat resistant
- test tubes, sterile (4)
- test-tube rack
- rubber stoppers, sterile (4)
- test-tube holder
- water bath, boiling

Procedure

PREPARE THE LAB MATERIALS

1. Work with a partner. Review all safety procedures, including sterile techniques and safe handling of microbe cultures, with your teacher.

2. Put on a lab apron, safety goggles, and disposable gloves.

3. **CAUTION: Handle glass test tubes carefully.** Obtain four sterilized (free from bacteria or other microorganisms) test tubes and a test-tube rack. Use a wax pencil to label the tubes "A" through "D." Place the tubes in the test-tube rack.

Name _____ Class _____ Date _____

Microbe Growth *continued*

4. Fill tubes A, B, and C halfway with sterile nutrient broth solution.
5. Insert a sterile rubber stopper into the mouth of tube A. For now, leave the other broth-filled tubes unsealed.
6. Fill tube D halfway with sterilized water and add the tube to the test-tube rack.
7. Use the table below for your observations of microbe growth. Observe the appearance of the broth in all four test tubes, and record your observations in the column labeled "Day 1."

Observations of Microbe Growth

	Tube Contents	Day 1	Day 2	Day 3	Day 4	Day 5	Day 6	Day 7
Tube A	Sterile broth							
Tube B	Broth exposed to air							
Tube C	Exposed broth, then sterilized							
Tube D	Sterilized water, exposed to air							

8. Once you have recorded your observations, place the test-tube rack in an area where it will not be disturbed for 24 hours.

MAKE OBSERVATIONS ON DAY TWO

9. Put on a lab apron, safety goggles, and disposable gloves.
10. Observe the appearance of the solution in each of the four test tubes. Record your observations in the column labeled "Day 2."
11. **CAUTION: Review the safety procedures associated with heating and electrical safety.** Put on heat-resistant gloves. With your teacher's guidance, use a test-tube holder to move tube C into a boiling water bath. Keep the test tube in the bath for 10 minutes.

Name _____ Class _____ Date _____

Microbe Growth *continued*

12. Use the test tube holder to remove tube C from the water. Allow the tube to cool, and return it to the test-tube rack.

13. Insert a sterile rubber stopper into the mouth of tubes B, C, and D.

14. Place the rack with all four sealed test tubes in an area where it will not be disturbed for several days.

MAKE CONTINUED OBSERVATIONS

15. Each day for the next five days, put on safety goggles and gloves. Carefully examine the contents of each tube. (Note: Make sure that the stopper remains in place, sealing the tube's opening.)

16. Write down your observations of the broth. Note any changes in the clarity of the solution. Also, look for changes in the appearance of the broth surface.

17. Clean up your lab materials according to your teacher's instruction. Wash your hands before you leave the lab.

Analyze and Conclude

1. **Summarizing Data** Did you observe any changes in the appearance of the tube contents over time? Describe your results for each of the four test tubes.

   _____
   _____
   _____
   _____
   _____
   _____
   _____

2. **Use Evidence to Make Explanations** If any of the tubes remained unchanged, explain why they did not support microbe growth.

   _____
   _____
   _____
   _____

Microbe Growth *continued*

3. **Scientific Methods** **Formulating Scientific Questions** Why was keeping all of the tubes sealed throughout the experiment essential?

___

Extension

4. **Further Inquiry** Does temperature affect microbe growth? How could you find out? Based on this investigation, design an experiment that would uncover any relationship between microbe growth and temperature. Share your experimental design with your teacher. With your teacher's permission, perform the investigation.

___

Name _____ Class _____ Date _____

Inquiry

DATASHEET B FOR IN-TEXT LAB

Enzymes in Detergents

OBJECTIVES

- **Recognize** the function of enzymes in laundry detergents.
- **Relate** temperature and pH to the activity of enzymes.

MATERIALS

- lab apron, safety goggles, and disposable gloves
- graduated cylinder
- beaker, 150 mL
- $NaHCO_3$ (0.1g)
- boiling water, 50 mL
- pH paper
- test-tube rack
- plastic wrap
- beakers, 50 mL (6)
- detergent, laundry, powdered, five brands (1 g of each)
- balance
- stirring rod, glass
- gelatin, instant, regular (18 g) or sugar-free (1.8 g)
- tongs or a hot mitt
- thermometer
- test tubes (6)
- pipet with bulb
- tape
- distilled water, 50 mL
- wax pencil
- metric ruler

Preparation

PREPARE THE LAB MATERIALS

1. **Scientific Methods State the Problem** How can you determine if a detergent contains proteases?

2. **Scientific Methods Form a Hypothesis** Form a testable hypothesis that explains how a protein mixture might be affected by a detergent that contains protease.

_____

_____

Name _____ Class _____ Date _____

Enzymes in Detergents *continued*

Procedure
MAKE A PROTEIN SUBSTRATE

1. Put on a lab apron, safety goggles, and gloves.

2. **CAUTION: Use tongs or a hot mitt to handle heated glassware.** Put 18 g of regular gelatin in a 150 mL beaker. Slowly add 50 mL of boiling water to the beaker, and stir the mixture with a stirring rod. Test and record the pH of this solution.

   _____

3. Very slowly add 0.1 g of $NaHCO_3$, baking soda, to the hot gelatin while stirring. Note any reaction. Test and record the pH of this solution.

   _____

4. Place six test tubes in a test-tube rack. Pour 5 mL of the gelatin-$NaHCO_3$ mixture into each tube. Use a pipet to remove any bubbles from the surface of the mixture in each tube. Cover the tubes tightly with plastic wrap and tape. Cool the tubes, and store them at room temperature until you begin your experiment.

5. Clean up your lab materials according to your teacher's instructions. Wash your hands before you leave the lab.

DESIGN AN EXPERIMENT

6. Design an experiment that tests your hypothesis and that uses the materials listed for this lab. Predict what will happen during your experiment if your hypothesis is supported.

   _____
   _____
   _____
   _____
   _____

Name _____ Class _____ Date _____

Enzymes in Detergents *continue*

7. Write a procedure for your experiment. Identify the variables that you will control, the experimental variables, and the responding variables. Construct any tables you will need to record your data. Make a list of all safety precautions you will take. Have your teacher approve your procedure before you begin the experiment.

_____

_____

_____

_____

CONDUCT YOUR EXPERIMENT

8. Put on a lab apron, safety goggles, and gloves.

9. Make a 10% msolution of each laundry detergent by dissolving 1 g of detergent in 9 mL of distilled water.

10. Carry out your experiment. Observe the test tubes after 24 hours. Record your data in a data table.

11. Clean up your lab materials according to your teacher's instructions. Wash your hands before you leave the lab.

Analyze and Conclude

1. **Scientific Methods Analyzing Methods** Suggest a reason for adding $NaHCO_3$ to the gelatin solution.

_____

_____

2. **Scientific Methods Summarizing Data** Make a bar graph of your data. Plot the amount of gelatin broken down (change in the depth of the gelatin) on the *y*-axis. Plot detergent on the *x*-axis.

3. **Scientific Methods Inferring Conclusions** What conclusions did your group infer from the results? Explain.

_____

_____

_____

_____

Name _____ Class _____ Date _____

Enzymes in Detergents *continued*

4. **Designing an Experiment** How can you test the effect of pH and temperature on action of enzymes in detergents?

_____
_____
_____

5. **Further Inquiry** Write a new question about enzymes in detergents that could be explored in another investigation.

_____
_____
_____
_____

Extensions

6. **Relating Concepts** What other active ingredients are present in laundry detergents, and how do they help clean clothes?

_____
_____
_____
_____

7. **Applying Information** What other household products contain enzymes, and what types of enzymes do they contain?

_____
_____
_____
_____

Name _____ Class _____ Date _____

Inquiry Lab

DATASHEET B FOR IN-TEXT LAB

Ecosystem Change

OBJECTIVES

- **Construct** an ecosystem model.
- **Observe** interactions of organisms in an ecosystem model.
- **Compare** an ecosystem model with a natural ecosystem.

MATERIALS

- goggles, gloves, and a lab apron
- terrarium or glass jar, large, with a lid
- grass seeds, a pinch of
- water, 150 mL
- mealworms (beetle larvae)
- earthworms
- crickets
- coarse sand or pea gravel
- soil
- clover seeds, a pinch of
- rolled oats
- mung bean seeds
- isopods (pill bugs)

Preparation

1. **State the Problem** How might the different organisms interact in an ecosystem model?

   _____
   _____
   _____
   _____

Name _____ Class _____ Date _____

Ecosystem Change *Continued*

2. **Form a Hypothesis** Form a testable hypothesis about how the number of individuals of each species in an ecosystem model will change over time.

_____

_____

_____

Procedure

BUILD AN ECOSYSTEM IN A JAR

1. **CAUTION: Glassware is fragile. Notify your teacher promptly of any broken glass or cuts. Do not clean up broken glass or spills that contain broken glass unless your teacher tells you to do so.** Place 5 cm of sand or pea gravel in the bottom of a large, clean glass jar that has a lid. Cover the gravel with 5 cm of soil.

2. Sprinkle the seeds of two or three kinds of small plants, such as grasses and clovers, onto the surface of the soil. Add about 150 mL of water. Put the lid on the jar loosely, and place the jar in indirect sunlight. Let the jar remain undisturbed for one week.

3. **CAUTION: Handle animals carefully.** Do not allow animals to escape from containers. After one week, place a handful of rolled oats into the jar. Place the mealworms in the oats. Then, place the other animals into the jar, and replace the lid. Place the lid on the jar loosely so that air can enter the jar.

DESIGN AN EXPERIMENT

4. Work with the members of your lab group to design an experiment that will test the hypothesis that you recorded previously. Design your experiment to use the ecosystem model you built.

5. Write a procedure for your experiment. Make a list of all the safety precautions that you will take. Have your teacher approve your procedure and safety precautions before you begin the experiment.

_____

_____

_____

_____

Name _____ Class _____ Date _____
Ecosystem Change *Continued*

6. Set up your group's experiment. Conduct your experiment for at least 14 days.

CLEANUP AND DISPOSAL

7.  Dispose of solutions, broken glass, and other materials in the designated waste containers. Do not put lab materials in the trash unless your teacher tells you to do so.

8. Clean up your lab materials according to your teacher's instructions. Wash your hands before you leave the lab.

Name _____ Class _____ Date _____

Ecosystem Change *Continued*

Analyze and Conclude

1. **Summarizing Results** Make graphs showing how the number of individuals of each species in your ecosystem changed over time. Plot time on the *x*-axis and the number of organisms on the *y*-axis.

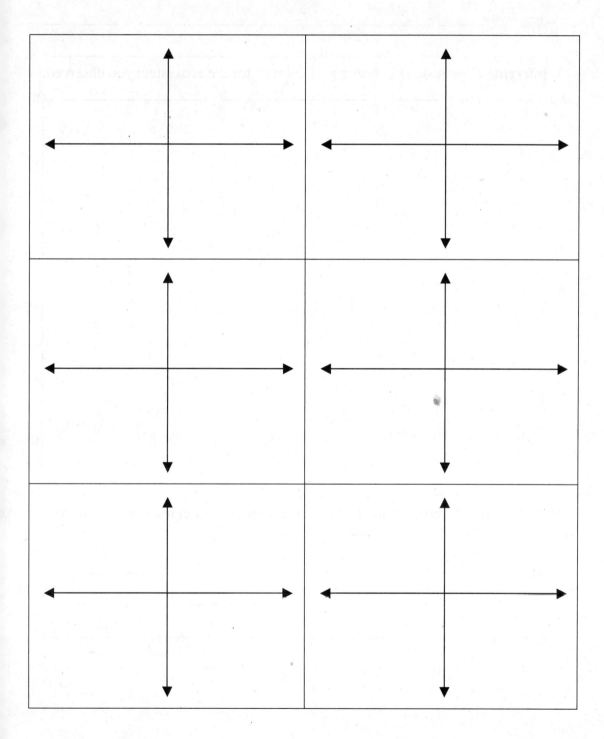

Holt Biology · 23 · Ecosystems

Name _____ Class _____ Date _____
Ecosystem Change *Continued*

2. **Analyzing Data** Compare your results with your hypothesis. Explain any differences.

_____

_____

_____

3. **Inferring Conclusions** Construct a food web for the ecosystem you observed.

[blank box for food web diagram]

4. **Recognizing Relationships** Does your ecosystem model resemble a natural ecosystem? Explain your answer.

_____

_____

_____

_____

Name _____ Class _____ Date _____

Ecosystem Change *Continued*

5. **Analyzing Methods** How can you build an ecosystem model that better represents a natural ecosystem?

_____
_____
_____
_____

6. **Critiquing Models** Was your ecosystem model truly a closed ecosystem? List your model's strengths and weaknesses as a closed ecosystem.

_____
_____
_____
_____

7. **Analyzing Data** List the biotic and abiotic factors in your ecosystem model.

_____
_____
_____
_____

Extensions

8. **Further Inquiry** Write a new question to explore with another investigation using an ecosystem model.

_____
_____
_____
_____

Name _____ Class _____ Date _____

Ecosystem Change *Continued*

9. **Making Comparisons** Use the library or Internet to learn about Biosphere 2. What problems did the Biosphere 2 crew encounter during the 1991–1993 project?

_____

_____

_____

_____

_____

_____

_____

_____

Name _____ Class _____ Date _____

Skills Practice DATASHEET B FOR IN-TEXT LAB

Yeast Population Growth

OBJECTIVES

- **Observe** the growth and decline of a population of yeast cells.
- **Determine** the carrying capacity of a yeast culture.

MATERIALS

- lab apron, safety goggles, and gloves
- test tubes (2)
- methylene blue solution, 1%
- coverslip
- yeast cell culture
- pipets, 1 mL (2)
- microscope slide, ruled
- microscope, compound

Procedure

COLLECTING DATA

1. **CAUTION: Do not touch or taste any chemicals. Know the location of the emergency shower and eyewash station and how to use them. Methylene blue will stain your skin and clothing.** Transfer 1 mL of yeast culture to a test tube. Add two drops of methylene blue to the test tube. The methylene blue will remain blue in dead cells but will turn colorless in living cells.

2. Make a wet mount by placing 0.1 mL, or about one drop, of the yeast culture and methylene blue mixture on a ruled microscope slide. Cover the slide with a coverslip.

3. Observe the wet mount under low power of a compound microscope. Notice the squares on the slide. Then switch to high power. (Note: Adjust the light so that you can clearly see both stained and unstained cells.) Move the slide so that the top left-hand corner of one square is in the center of your field of view. This area will be area 1, as shown in the diagram.

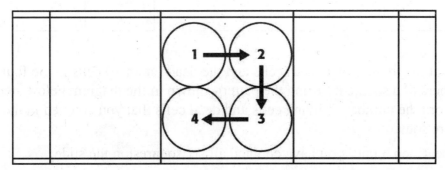

Name _____ Class _____ Date _____
Yeast Population Growth *continued*

4. Use the two data tables below to record your observations of living and dead cells.

Number of Living Cells

Time (h)	1	2	3	4	5	6	Average
0							
24							
48							
72							
96							

Number of Dead Cells

Time (h)	1	2	3	4	5	6	Average
0							
24							
48							
72							
96							

5. Count the living (unstained) cells and the dead (stained) cells in the four corners of a square by using the pattern shown in the diagram below step 3. Record the number of living cells and dead cells that you counted in the entire square.

6. Repeat step 5 until you have counted all six squares on the slide.

Name _____ Class _____ Date _____

Yeast Population Growth *continued*

7.  Clean up your lab materials according to your teacher's instruction. Wash your hands before you leave the lab.

COMPILING DATA

8. Refer to your first data table. Find the total number of living cells in the six squares. Divide this total by 6 to find the average number of living cells per square. Record this number in your data table. Repeat this procedure for the dead cells.

9. Repeat steps 1 through 5 each day for four more days.

Analyze and Conclude

1. **Evaluating Methods** Explain why several areas were counted and averaged each day.

   _____

   _____

   _____

2. **Analyzing Data** Graph the changes in the numbers of living yeast cells and dead yeast cells over time. Plot the number of cells in 1 mL of yeast culture on the *y*-axis and the time (in hours) on the *x*-axis.

3. **Evaluating Results** Describe the general population changes that you observed in the yeast cultures over time.

   _____

   _____

   _____

4. **Scientific Methods Inferring Conclusions** Did the yeast population appear to reach a certain carrying capacity? What limiting factors probably caused the yeast population to decline?

   _____

   _____

   _____

   _____

Name _____ Class _____ Date _____

Yeast Population Growth *continued*

Extensions

5. **Designing an Investigation** Write a question about population growth that could be explored in another investigation. Design an investigation that could help answer that question.

_____
_____
_____
_____
_____
_____
_____
_____
_____

Name _____ Class _____ Date _____

Inquiry Lab

DATASHEET B FOR IN-TEXT LAB

Effects of Acid Rain on Seeds

OBJECTIVES

- **Simulate** an environmental condition in the laboratory.
- **Measure** the difference between treated and untreated seedlings.
- **Analyze** the effects of acidic conditions on plants.

MATERIALS

- beaker (250 mL)
- water, distilled
- solutions of various pH
- bags, plastic, resealable
- graph paper
- seeds (50)
- mold inhibitor (20 mL)
- paper towels
- pencil, wax (or marker)
- metric ruler

Preparation

1. **Scientific Methods State the Problem** How does acid rain affect plants?
2. **Scientific Methods Form a Hypothesis** Form a testable hypothesis that explains how a germinating plant might be affected by acid rain. Record your hypothesis.

_____

_____

_____

_____

Name _____ Class _____ Date _____
Effects of Acid Rain on Seeds *continued*

Procedure
DESIGN AN EXPERIMENT

1. Design an experiment that tests your hypothesis and that uses the materials listed for this lab. Predict what will happen during your experiment if your hypothesis is supported.

   _____
   _____
   _____

2. Write a procedure for your experiment. Identify the variables that you will control, the experimental variables, and the responding variables. Make a list of all of the safety precautions that you will take. Have your teacher approve your procedure before you begin.

   _____
   _____
   _____
   _____
   _____

CONDUCT YOUR EXPERIMENT

3. Put on safety goggles, gloves, and a lab apron.

4. **CAUTION: The mold inhibitor contains household bleach, which is a toxic chemical and a base.** Place your seeds in a 250 mL beaker, and slowly add enough mold inhibitor to cover the seeds. Soak the seeds for ten minutes, and then pour the mold inhibitor into the proper waste container. Gently rinse the seeds with distilled water, and place them on clean paper towels.

5. **CAUTION: Solutions that have a pH below 7.0 are acids.** Carry out your experiment for 7–10 days. Make observations every 1–2 days, and note any changes. Record your observations each day in the data table.

Name _____ Class _____ Date _____

Effects of Acid Rain on Seeds *continued*

Effects of Acid Rain

Solution	Date	Observations

Name _____ Class _____ Date _____

Effects of Acid Rain on Seeds *continued*

6. Clean up your lab materials according to your teacher's instructions. Wash your hands before leaving the lab.

Analyze and Conclude

1. **Summarizing Results** Describe any changes in the look of your seeds during the experiment. Discuss seed type, average seed size, number of germinated seeds, and changes in seedling length.

_____
_____
_____
_____
_____
_____
_____
_____
_____

2. **Analyzing Results** Were there any differences between the solutions? Explain.

_____
_____

3. **Analyzing Methods** What was the control group in your experiment?

_____

4. **Analyzing Data** Make graphs of your group's data. Plot seedling growth (in millimeters) on the *y*-axis. Plot number of days on the *x*-axis.

5. **Scientific Methods Interpreting Data** How do acidic conditions appear to affect seeds?

_____
_____

6. **Predicting Outcomes** How might acid rain affect the plants in an ecosystem?

_____
_____
_____

Name _____ Class _____ Date _____

Effects of Acid Rain on Seeds *continued*

7. **Scientific Methods** **Critiquing Procedures** How could your experiment be improved?

_____
_____
_____
_____

8. **Scientific Methods** **Formulating Scientific Questions** Write a new question about the effect of acid rain that could be explored with another investigation.

_____
_____
_____

Extensions

9. **Inferring Relationships** Research to identify the parts of the United States that are most affected by acid rain. Explain why acid rain affects these areas more than it affects other areas.

_____
_____
_____
_____
_____

10. **Analyzing Methods** Describe how factories have changed to reduce the amount of acid rain.

_____
_____
_____
_____
_____
_____

Name _____ Class _____ Date _____

Skills Practice Lab   DATASHEET B FOR IN-TEXT LAB

Plant Cell Observation

OBJECTIVES

- **Identify** the structures that you can see in plant cells.
- **Compare** and **contrast** the structures that you can see in stained plant cells.

MATERIALS

- compound light microscope
- *Elodea* sprig
- microscope slides solution and coverslips
- lamp, incandescent
- forceps
- dropper bottle of Lugol's iodine

Procedure

1. Put on safety goggles, gloves, and a lab apron.

2. **CAUTION: Handle glass slides and coverslips with care.** Using forceps, carefully remove a small leaf from the top of an *Elodea* sprig. Place the whole leaf in a drop of water on a slide. Add a coverslip.

3. Observe the leaf under low power. Look at an area where you can see the cells clearly. Switch to high power.

4. Focus on an *Elodea* cell in which you can see the chloroplasts clearly. Draw this cell. Label the cell wall, a chloroplast, and any other cell parts that you can see.

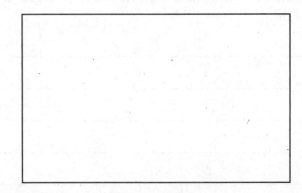

Name _____ Class _____ Date _____

Plant Cell Observation *continued*

5. Notice if the chloroplasts are moving in any of the cells. If you do not see movement, warm the slide in your hand or under a bright lamp for a minute or two. Look for movement of the cell contents—called *cytoplasmic streaming*—again under high power.

6. **CAUTION: Lugol's solution is toxic and stains skin and clothing. Promptly wash off spills.** Make a wet mount of another leaf from the elodea sprig, but use Lugol's iodine solution instead of water. Observe the cells of the leaf under low power and high power.

7. Draw a stained *Elodea* cell. Label the cell wall, a chloroplast, the central vacuole, the nucleus, and the cell membrane.

8. Clean up your lab materials according to your teacher's instruction. Wash your hands before you leave the lab.

Analyze and Conclude

1. **Analyzing Methods** What structures became visible when the *Elodea* cells were stained with iodine?

   _____

   _____

2. **Inferring Conclusions** Cytoplasmic streaming occurs only in living cells. What can you conclude about the effect of Lugol's iodine solution on plant cells?

   _____

   _____

   _____

Name _____ Class _____ Date _____

Inquiry

DATASHEET B FOR IN-TEXT LAB

Cell Size and Diffusion

OBJECTIVES
- **Relate** a cell's size to its surface area-to-volume ratio.
- **Predict** how the surface area-to-volume ratio of a cell will affect the diffusion of substances into the cell.

MATERIALS
- safety goggles
- disposable gloves
- knife, plastic
- beaker, 250 mL
- spoon, plastic
- lab apron
- block of phenolphthalein agar (3 cm × 3 cm × 6 cm)
- ruler, metric
- vinegar, 150 mL
- paper towel

Preparation

1. **Scientific Methods State the Problem** How does a cell's size affect the delivery of substances via diffusion to the center of the cell?

2. **Scientific Methods Form a Hypothesis** Form a testable hypothesis that explains how a cell's size affects the rate of diffusion of substances from outside the cell.

_____
_____

Procedure
DESIGN AN EXPERIMENT

1. Design an experiment that tests your hypothesis and that uses the materials listed for this lab. Predict what will happen during your experiment if your hypothesis is correct.

_____
_____
_____
_____

Name _____ Class _____ Date _____

Cell Size and Diffusion *continued*

2. Write a procedure for your experiment. Identify the variables that you will control, the experimental variables, and the responding variables. Construct any tables that you will need to record your data. Make a list of all safety precautions that you will take. Have your teacher approve your procedure before you begin.

_____

_____

_____

_____

_____

_____

Diffusion in Cubes		
Size (cm)	Ratio	Distance (mm)

CONDUCT YOUR EXPERIMENT

3. Put on safety goggles, gloves, and a lab apron.

4. Carry out your experiment. Record your observations in the data table on the previous sheet.

5. Clean up your lab materials according to your teacher's instruction. Wash your hands before you leave the lab.

Holt Biology — Cells and Their Environment

Name _____ Class _____ Date _____

Cell Size and Diffusion *continued*

Analyze and Conclude

1. **Interpreting Observations** Describe any changes in the appearance of the agar cubes. Explain why these changes occurred.

   _____

   _____

   _____

   _____

2. **Summarizing Results** Make a graph labeled "Diffusion distance (mm)" on the vertical axis and "Surface area-to-volume ratio" on the horizontal axis. Plot your group's data on the graph.

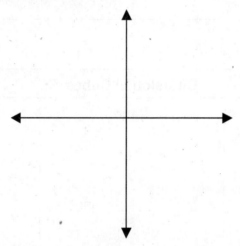

3. **Scientific Methods Analyzing Results** Using the graph you made in item 2, make a statement relating the surface area-to-volume ratio and the distance that the substance diffuses.

   _____

   _____

   _____

   _____

Name _____ Class _____ Date _____

Cell Size and Diffusion *continued*

4. **Summarizing Results** Make a second graph using your group's data. Label the vertical axis "Rate of diffusion (mm/min)" (distance that vinegar moved ÷ time). Label the horizontal axis "Surface area-to-volume ratio." Plot your group's data on the graph.

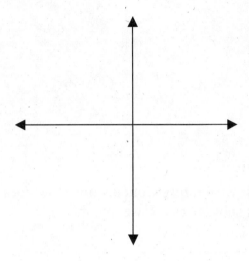

5. **Analyzing Results** Referring to the graph that you made in item 4, make a statement relating the surface area-to-volume ratio and the rate at which the substance diffuses.

_____
_____
_____
_____

6. **Scientific Methods Evaluating Methods** In what ways do your agar models simplify or fail to simulate the features of cells? (Hint: What processes are they incapable of reproducing?)

_____
_____
_____
_____

Name _____ Class _____ Date _____
Cell Size and Diffusion *continued*

7. **Calculating** Calculate the surface area and the volume of a cube that has a side length of 5 cm. Calculate the surface area and volume of a cube that has a side length of 10 cm. Determine the surface area-to-volume ratio of each cube. Which cube has the greater surface area-to-volume ratio?

8. **Scientific Methods** **Evaluating Conclusions** How does the size of a cell affect the rate at which substances diffuse into the cell?

   _____

   _____

9. **Further Inquiry** Write a new question about cell size and diffusion that could be explored in another investigation.

   _____

   _____

   _____

Extensions

10. How does cell transport in prokaryotic cells differ from cell transport in eukaryotic cells?

    _____

    _____

    _____

11. Which of the following can diffuse across the cell membrane without the help of a transport protein: water, carbohydrates, lipids, or proteins?

    _____

    _____

    _____

Name _____ Class _____ Date _____

Skills Practice

DATASHEET B FOR IN-TEXT LAB

Cellular Respiration

OBJECTIVES

- **Demonstrate** how carbon dioxide affects bromothymol blue when added to the indicator solution.
- **Describe** the effect of temperature on carbon dioxide production by yeast.

MATERIALS

- safety goggles
- lab apron
- room temperature water
- drinking straw, plastic
- ice water
- ¼ teaspoon
- sugar
- disposable gloves
- plastic cups, clear (4)
- bromothymol blue
- warm water
- baker's yeast
- hand lens

Procedure

1.  Put on safety goggles, gloves, and a lab apron. Fill a clean plastic cup halfway with room temperature water. Add several drops of bromothymol blue to the water. Swirl to mix the solution.

     CAUTION: Bromothymol blue is a skin and eye irritant.

2. Insert a clean straw into the solution. Gently blow a steady stream of air through the straw. Note any changes to the solution's appearance. **CAUTION: Be careful not to accidentally drink the solution while blowing into the straw.**

    _____

    _____

3. Label three plastic cups "A," "B" and "C."

Name _____ Class _____ Date _____

Cellular Respiration *continued*

4. Fill cup A with ice water, fill cup B with room temperature water, and fill cup C with warm water. Add several drops of bromothymol blue solution to each cup to ensure a uniform appearance.

5. Add ¼ teaspoon of baker's yeast to each cup. Swirl the cups, and observe the appearance of the solutions every 30 seconds. After 5 minutes, examine the surface of each solution with a hand lens.

6.  Clean up your lab materials according to your teacher's instruction. Wash your hands before you leave the lab.

Analyze and Conclude

1. **Drawing Conclusions** What happened to the indicator as exhaled air bubbled through the solution? What caused this change?

   _____

   _____

   _____

2. **Scientific Methods Evaluating Results** Did the yeast produce a similar color change? Explain your answer.

   _____

   _____

   _____

3. **Scientific Methods Evaluating Results** Did temperature affect the yeast's production of carbon dioxide? Explain your answer.

   _____

   _____

   _____

4. **Scientific Methods Summarizing Results** What did you observe on the surface of the solutions?

   _____

   _____

   _____

Cellular Respiration *continued*

5. **Predicting Outcomes** Will adding sugar to the yeast solution affect the respiration rate? Make a guess. Then, design a method for inquiry that would test the effects of various sugar concentrations on yeast metabolism.

Name _____ Class _____ Date _____

Skills Practice   DATASHEET B FOR IN-TEXT LAB

Mitosis in Plant Cells

OBJECTIVES

- **Examine** the dividing root-tip cells of an onion.
- **Identify** the phase of mitosis that each cell in an onion root tip is undergoing.
- **Determine** the relative length of time each phase of mitosis takes in onion root-tip cells.

MATERIALS

- compound light microscope
- prepared microscope slide of a longitudinal section of *Allium* (onion) root tip

Procedure

IDENTIFY THE PHASES OF MITOSIS

1. 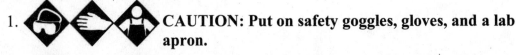 **CAUTION: Put on safety goggles, gloves, and a lab apron.**

2. 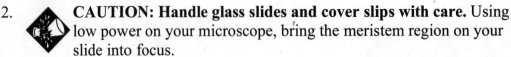 **CAUTION: Handle glass slides and cover slips with care.** Using low power on your microscope, bring the meristem region on your slide into focus.

3. Examine the meristem carefully. Choose a sample of about 50 cells. Look for a group of cells that appear to have been actively dividing at the time that the slide was made. The cells will appear to be in rows, so it should be easy to keep track of them. The dark-staining bodies are the chromosomes.

4. For each of the cells in your sample, identify the stage of mitosis. Use the data table on the next page to show the relative duration of each phase of mitosis. Record your observations in the data table.

Name _____ Class _____ Date _____
Mitosis in Plant Cells *continued*

Relative Duration of Each Phase of Mitosis

Phase of mitosis	Tally marks	Count	Percentage of all cells	Time (min)
Prophase				
Metaphase				
Anaphase				
Telophase				

CALCULATE THE RELATIVE LENGTH OF EACH PHASE

5. When you have classified each cell in your sample, count the tally marks for each phase and fill in the "Count" column.

   In which phase of mitosis was the number of cells the greatest? _____

   In which phase of mitosis was the number of cells the fewest? _____

6. Calculate what percentage of all cells were found in each phase. Divide the number of cells in a phase by the total number of cells in your sample, and multiply by 100. Enter these figures under the "Percentage" column.

$$\text{Percentage} = \frac{\text{number of cells in phase}}{\text{total number of cells in sample}} \times 100\%$$

7. The percentage of the total number of cells that are found in each phase can be used to estimate how long each phase lasts. For example, if 25% of the cells are in prophase, then prophase takes 25% of the total time that a cell takes to undergo mitosis. Mitosis in onion cells takes about 80 min. Using this information and the percentages you have just determined, calculate the time for each phase and record it in your data table.

$$\text{Duration of phase (in minutes)} = \frac{\text{percentage}}{100} \times 80 \text{ min}$$

8. Use the table on the following page to record the data for the entire class. Collect and add the counts for each phase of mitosis for the entire class. Fill in the percentage and time information by using these data.

Name _____ Class _____ Date _____

Mitosis in Plant Cells *continued*

Class Data

Phase of mitosis	Count	Percentage of all cells	Duration (min)
Prophase			
Metaphase			
Anaphase			
Telophase			

9. Clean up your lab materials according to your teacher's instructions. Wash your hands before leaving the lab.

Analyze and Conclude

1. **Identifying Structures** What color are the chromosomes stained?

   _____

2. **Recognizing Relationships** How can you distinguish between early and late anaphase?

   _____

   _____

3. **Scientific Methods Making Systematic Observations** According to your data table, which phase takes the least amount of time? _____

   Which phase of mitosis lasts the longest? _____

   Why might this phase require more time than other phases of mitosis do?

   _____

   _____

   _____

Mitosis in Plant Cells *continued*

4. **Scientific Methods Summarizing Data** How does your data compare with the data of the entire class?

_____

_____

5. **Scientific Methods Critiquing Procedures** In this investigation, you assumed that the percentage of total time that any given phase takes is equal to the percentage of the total number of cells in that phase at any moment. Why might this not be true for very small samples of cells?

_____

_____

_____

_____

Extensions

6. **Applying Methods** Cancerous tissue is composed of cells undergoing uncontrolled, rapid cell division. How could you develop a procedure to identify cancerous tissue by counting the number of cells undergoing mitosis?

_____

_____

_____

_____

_____

_____

Name _____ Class _____ Date _____

Skills Practice

DATASHEET B FOR IN-TEXT LAB

Meiosis Model

OBJECTIVES

- **Model** the stages of meiosis.
- **Describe** the events that occur in each stage of the process of meiosis.
- **Compare** your meiosis model to meiosis stages in a set of prepared slides of lily anther microsporocytes.

MATERIALS

- beads, wooden (40)
- marker
- microscope slides of lilium anther, 1st and 2nd meiotic division
- tape, masking
- index cards (8)
- microscope
- scissors
- yarn

Procedure

BUILD A MODEL

1. Work in a team of two. Review the stages of meiosis I and meiosis II. Note the structures and organization that are characteristic of each stage. Pay particular attention to the appearance and behavior of the chromosomes.

2. Work with your partner to design a model of a cell by using the materials listed for this lab. Select and assign a different material to represent each cell structure and keep this consistent in all models. Have your teacher approve the plan.

3. Label each of the eight index cards with a specific stage of meiosis, such as "Prophase II."

4. Using your model plan that you designed in step 2, you or your partner will construct a set of models representing the four stages of meiosis I. The other team member will construct another set of models representing the four stages of meiosis II.

Name _____ Class _____ Date _____

Meiosis Model *continued*

5. Once you have completed your set of models, position the cards in two horizontal rows. The top row illustrates the stages of meiosis I. The bottom row illustrates the stages of meiosis II. Compare and contrast the corresponding stages.

_____
_____
_____
_____

OBSERVE MEIOSIS

6. **CAUTION: Handle glass slides with care.** Get a set of prepared slides of lily anther microsporocytes that show different meiotic stages.

7. Use your microscope to view each slide. Locate the various stages of meiosis within the anther sacs.

8. Compare what you observe in the prepared slides to the models that you have constructed.

_____
_____
_____
_____

Analyze and Conclude

1. **Analyzing Processes** Identify and label each stage of meiosis as a haploid stage or a diploid stage.

_____
_____
_____
_____

2. **Comparing Functions** How does anaphase I differ from anaphase II?

_____
_____
_____
_____

Name _____ Class _____ Date _____

Meiosis Model *continued*

3. **Scientific Methods Critiquing Models** Based upon the observations of real cells, evaluate your model. How would you improve your model?

_____

_____

_____

_____

_____

_____

Name _____ Class _____ Date _____

Inquiry Lab

DATASHEET B FOR IN-TEXT LAB

Plant Genetics

OBJECTIVES

- **Develop** a hypothesis to predict the yield of a corn crop.
- **Design** and conduct an experiment to test your hypothesis.
- **Compare** germination and survival rates of three lots of corn seeds.

MATERIALS

- lab apron, disposable gloves
- corn seeds, normal (10 from lot A and 10 from lot B)
- corn seeds, 3:1 mix of normal and albino (10 from lot C)
- plant tray or pots
- soil, potting (3 kg)
- water

Preparation

1. **State the Problem** What might happen to a seed that has one or more albinism alleles?

   _____

   _____

2. **Form a Hypothesis** Form a hypothesis about how albinism affects the success of plants grown from seed.

   _____

   _____

Procedure

DESIGN AN EXPERIMENT

1. Design an experiment that will determine the germination and survival rates of three lots of corn seeds of unknown genotype. Write out a procedure for your experiment on a separate sheet of paper. Be sure to include safety procedures, and construct tables to organize your data. Have your teacher approve your plan before you begin.

2. Predict the outcome of your experiment, and record this prediction.

Plant Genetics *continued*

CONDUCT YOUR EXPERIMENT

3.  **CAUTION: Wear gloves and a lab apron whenever handling soil, seeds, or plants.**

4. Follow your written procedure. Make note of any changes.

5. Record all data in your tables. Also record any other observations.

6. At the end of the experiment, present your results to the class. Devise a way to collect the class data in a common format.

7. Clean up your lab materials according to your teacher's instruction. Wash your hands before you leave the lab.

Analyze and Conclude

1. **Evaluating Experimental Design** Did you get clear results? How might you improve your design?

   _____

   _____

   _____

   _____

2. **Analyzing Results** Did your results support your hypothesis? Explain your answer.

   _____

   _____

   _____

   _____

3. **Analyzing Data** Use the class data to calculate the average germination rate and survival rate for each lot of corn seeds. Describe any patterns that you notice.

   _____

   _____

   _____

   _____

   _____

Name _____ Class _____ Date _____

Skills Practice Lab   DATASHEET B FOR IN-TEXT LAB

DNA Extraction from Wheat Germ

OBJECTIVES

- **Extract** DNA from wheat germ.
- **Explain** the role of detergents, heat, and alcohol in the extraction of DNA.

MATERIALS

- test tube or beaker (50 mL)
- salt, table
- isopropyl alcohol, cold (15 mL)
- inoculating loop
- wheat germ, raw (1 g)
- water, hot tap (55 C, 20 mL)
- soap, liquid dishwashing (1 mL)
- glass rod, 8 cm long
- glass slide

Procedure

1. Put on safety goggles, lab apron, and gloves.

2. **CAUTION: Glassware, such as a test tube, is fragile and can break.** Place 1 g of wheat germ into a clean test tube.

3. Add 20 mL hot (55°C) tap water and stir with glass rod for 2 to 3 min.

4. Next, add a pinch of table salt, and mix well.

5. Add a few drops (1 mL) of liquid dishwashing soap. Stir the mixture with the glass rod for 1 min until it is well mixed.

6. **CAUTION: Isopropyl alcohol is flammable. Bunsen burners and hot plates should be removed from the lab.** Slowly pour 15 mL cold isopropyl alcohol down the side of the tilted tube or beaker. The alcohol should form a top layer over the original solution. Note: Do not pour the alcohol too fast or directly into the wheat germ solution.

7. Tilt the tube upright, and watch the stringy, white material float up into the alcohol layer (this result should occur after 10 to 15 min). This material is the DNA from the wheat germ.

Name _____ Class _____ Date _____

DNA Extraction from Wheat Germ *continued*

8. Carefully insert the inoculating loop into the white material in the alcohol layer. Gently twist the loop as you wind the DNA around the loop. Remove the loop from the tube, and tap the DNA onto a glass slide.

9. Clean up your lab materials according to your teacher's instruction. Wash your hands before you leave the lab.

Analyze and Conclude

1. **Describing Events** Describe the appearance of the DNA on the slide.

   _____
   _____

2. **Interpreting Information** Explain the role of detergent, heat, and isopropyl alcohol in the extraction of DNA.

   _____
   _____
   _____
   _____

3. **Scientific Methods Comparing Structures** How do the characteristics of your DNA sample relate to the structure of eukaryotic DNA?

   _____
   _____

4. **Scientific Methods Designing Experiments** Design a DNA extraction experiment in which you explore the effect of changing the variables.

   _____
   _____
   _____
   _____
   _____
   _____
   _____

Name _____ Class _____ Date _____

Skills Practice

DATASHEET B FOR IN-TEXT LAB

Protein Detection

OBJECTIVES

- **Perform** a protein assay to detect the results of gene expression.
- **Use** gel electrophoresis and staining to detect size differences.
- **Infer** the presence of similar genes in different species.

MATERIALS

- lab apron, safety goggles, and disposable gloves
- microtubes, flip-top or screw-top (6 to 12)
- precast gel for electrophoresis chamber
- electrophoresis chamber with power supply and wires
- protein stain solution
- fish muscle samples, (3 to 6 unknowns) electrophoresis chamber
- protein buffer solution with dye
- water bath
- micropipettes or tips, sterile, disposable (3 to 6)
- running buffer solution
- gel staining tray
- water, distilled

Procedure

PREPARE PROTEIN SAMPLES

1.  Put on a lab apron, safety goggles, and gloves. Read all procedures, and prepare to collect your data. For each sample of fish muscle, record the type of fish, and assign it a code letter. Then, mark the letter onto two microtubes.

2. **CAUTION: Never eat or taste food in the lab.** Obtain a small piece of each fish muscle sample. Place each piece in a microtube that has the correct code label.

3. **CAUTION: Never taste chemicals or allow them to contact your skin.** For each sample, add enough protein buffer solution to cover the sample piece. Cap the tube, then gently flick it to mix the contents. The buffer will cause some of the proteins from each fish muscle sample to become suspended in the solution.

4. Let the tubes sit at room temperature for 5 min. Then, pour just the liquid from each into the second tube with the matching label. Keep the samples on ice until used.

Protein Detection continued

5. **CAUTION: Use extreme caution when working with heating devices.** Heat the samples in the water bath at 95 °C for 5 min. The heat will cause the proteins from each fish muscle to denature.

SEPARATE PROTEINS BY GEL ELECTROPHORESIS

6. Examine the gel that is precast within its chamber. Note the row of small wells along one edge. These wells are where you will place the samples to be separated. Keep the gel level as you work.

7. Slowly add running buffer solution to the chamber. Add just enough to flood the wells and to cover the gel surface with buffer about 2 mm deep. Be careful not to damage the gel while pouring.

8. Using a clean micropipette, transfer 10 µL of one sample solution into one well. Be careful not to overflow the well or puncture the gel with the pipette tip. Record the sample's "lane" position.

9. Repeat step 8 for each sample. Be sure to use a clean pipette for each transfer.

10. **CAUTION: Use caution when working with electrical equipment.** Assemble the electrophoresis chamber and power source as directed by your teacher. With your teacher's approval, connect the power supply to the chamber electrodes. The negative terminal should be connected to the electrode closest to the wells, and the positive terminal should be connected to the opposite electrode.

11. Leave the chamber running but undisturbed for the amount of time specified by your teacher. During this time, the samples and dye should move toward the positive side of the gel.

VIEW THE SEPARATED PROTEINS

12. When the moving front of the dye has migrated across the entire gel, disconnect the electrodes from the power source. Gently transfer the gel to the staining tray.

13. **CAUTION: Dispose of materials as directed by your teacher.** Gently pour off the buffer solution into an appropriate container as directed by your teacher.

14. Slowly pour the protein stain solution over the staining tray, and then wait for the amount of time specified by your teacher.

Protein Detection *continued*

15. Destain the gel by soaking and rinsing it several times in distilled water. Dispose of the rinse water as directed by your teacher. Some of the stain will remain on the proteins in the gel. Draw, photograph, or photocopy the gel for analysis.

16.  Clean up and dispose of your lab materials and waste according to your teacher's instructions. Wash your hands before leaving the lab.

Analyze and Conclude

1. **Scientific Methods** **Organizing Data** On your picture of the gel, mark the position of each visible band in each lane of the unknown samples. Also mark the bands in the lane(s) of the standard(s).

2. **Scientific Methods** **Analyzing Data** Compare the numbers and positions of visible bands among all lanes. Identify which bands of the unknown samples match bands of the standard(s). Also, identify which bands of the unknown samples match each other.

Protein Detection *continued*

Extensions

3. **Cladistic Analysis** Use the number of similarities between the samples to construct a cladogram (as explained in an earlier chapter). First, make a table to analyze each similarity. Then, draw a simple cladogram.

Name _____ Class _____ Date _____

Skills Practice

DATASHEET B FOR IN-TEXT LAB

DNA Fingerprint Analysis

OBJECTIVES

- **Model** the forensic analysis of evidence from a crime scene.
- **Use** restriction enzymes, PCR, and gel electrophoresis to manipulate DNA samples.
- **Compare** DNA fingerprints to match identical DNA samples.

MATERIALS

- lab apron, safety goggles, and disposable gloves
- microcentrifuge tubes (5)
- DNA samples (5)
- restriction enzyme
- ice, crushed
- gel, agarose, precast for electrophoresis chamber
- running buffer
- bag, plastic, resealable
- tray for staining gel
- marker, permanent, waterproof
- micropipettes, sterile, disposable (25)
- restriction enzyme buffer
- incubator or hot water bath
- cup, plastic-foam
- electrophoresis chamber with power supply and wires
- loading dye
- DNA staining solution
- water, distilled
- paper, white, or light table

Procedure

CUT DNA WITH RESTRICTION ENZYME

1. Read all procedures, and prepare to collect your data. Label each microcentrifuge tube with a code for each DNA sample provided. For example, label one tube "C" for "crime scene sample" and the remaining tubes "S1" to "S4," one for each suspect.

2.  Wear a lab apron, safety goggles, and gloves during all parts of this lab.

DNA Fingerprint Analysis continued

3. **CAUTION: Never taste chemicals or allow them to touch your skin.** Using a clean pipette each time, transfer 10 μL of each DNA sample to the microcentrifuge tube that has the matching label.

4. Using a clean pipette each time, transfer 2 μL of restriction enzyme buffer to each of the tubes.

5. Using a clean pipette each time, transfer 2 μL of restriction enzyme to each of the tubes. Close all of the tubes. Gently flick the bottom of each tube to mix the DNA and reagents.

6. **CAUTION: Use caution when working with heating devices.** Transfer the tubes to the incubator or water bath set at 37°C. Let the samples incubate for one hour.

7. Stand the tubes in crushed ice in the plastic-foam cup.

8. If you need to pause the lab at this point, store the cup at 4°C.

SEPARATE FRAGMENTS BY GEL ELECTROPHORESIS

9. Place the precast gel on the level surface of the electrophoresis chamber. The wells in the gel should be closest to the black, or negative, electrode. Keep the gel level and flat at all times.

10. Fill the chamber with enough buffer to barely cover the gel. Do not pour the buffer directly onto the gel. Sketch a diagram of your gel in the space below.

Lane 1	Lane 2	Lane 3	Lane 4	Lane 5
Crime scene DNA	Suspect #1 DNA	Suspect #2 DNA	Suspect #3 DNA	Suspect #4 DNA

Name _____ Class _____ Date _____

DNA Fingerprint Analysis *continued*

11. Using a clean pipette each time, transfer 2 μL of loading dye to each of the tubes. Gently flick the tubes to mix the contents.

12. Using a clean pipette, load the crime scene DNA into the well for Lane 1 of your gel. Be careful not to overflow or puncture the well.

13. Repeat step 12 for the remaining DNA samples and gel lanes. End with the DNA from suspect 4. Use a clean pipette for each transfer.

14. **CAUTION: Use caution when working with electrical equipment; use only as directed by your teacher.** Make sure that everything outside the chamber is dry before proceeding. Attach the power connectors to the chamber and power supply as directed by your teacher. Set the power supply to the voltage determined by your teacher, and turn on the power supply.

15. Allow the gel to run undisturbed for the time directed by your teacher. Observe the gel periodically, and stop the process when the dye front is about 3 cm away from the end of the gel. At that point, turn off the power supply. Then, disconnect the power connectors from the power supply and chamber.

16. **CAUTION: Dispose of all waste materials as directed by your teacher.** Carefully remove the casting tray from the chamber. Pour off the running buffer according to your teacher's instructions.

17. If you need to pause this lab at this point, carefully slide the gel into a resealable bag. Add 2 mL running buffer, seal the bag, and store the bag in a refrigerator. Remember to keep the gel flat.

VIEW SEPARATED DNA FRAGMENTS

18. Gently slide the gel onto the staining tray. Pour enough stain into the tray to barely cover the gel. Do not pour the stain directly onto the gel. Let the gel sit for at least 30 min.

19. Carefully pour off the stain as directed by your teacher.

20. Gently pour distilled water into the tray to cover the gel. Do not pour the water directly onto the gel. After 5 min, carefully pour off the water as directed by your teacher.

21. Repeat step 20 until bands are clearly visible on the gel.

22. Gently transfer the gel to a white sheet of paper or to a light table. Sketch and describe your observations on the top of the next page.

Name _____ Class _____ Date _____

DNA Fingerprint Analysis *continued*

[blank box for sketch]

_____
_____
_____
_____

23. Clean up your lab materials according to your teacher's instructions. Wash your hands before leaving the lab.

Analyze and Conclude

1. **Scientific Methods Organizing Data** Organize your data into a table on a separate sheet. How many different fragment sizes resulted from the treatment of each DNA sample?

_____
_____

2. **Analyzing Data** Identify any bands of fragments that are the same size among any of the samples. Mark these bands on your sketch.

Name _____ Class _____ Date _____
DNA Fingerprint Analysis *continued*

3. **Forming Conclusions** Use this evidence to determine which suspect most likely committed the crime. Explain your answer.

   _____

   _____

   _____

4. **Scientific Methods Evaluating Methods** Do these results provide enough evidence to convict the suspect? Explain your answer.

   _____

   _____

   _____

   _____

Extension

5. **Applying Concepts** Some bands appear in the same position in several lanes. Propose an explanation for this result.

   _____

   _____

   _____

   _____

6. **Predicting Results** How might the results have been affected if a different restriction enzyme had been used? Explain your answer.

   _____

   _____

   _____

   _____

Name _____ Class _____ Date _____

Skills Practice Lab DATASHEET B FOR IN-TEXT LAB

Natural Selection Simulation

OBJECTIVES

- **Model** natural selection.
- **Relate** favorable mutations to selection and evolution.

MATERIALS

- construction paper
- scissors
- soda straws
- penny or other coin
- meterstick or tape measure
- cellophane tape
- marker, felt-tip
- die, six-sided

Procedure

MODEL PARENTAL GENERATION

1. Cut a sheet of paper into two strips that are 2 cm × 20 cm each. Make a loop with one strip of paper. Let the paper overlap by 1 cm, and tape the loop closed. Repeat for the other strip.

2. Tape one loop 3 cm from one end of the straw and one loop 3 cm from the other end, as pictured. Use a felt-tip marker to mark the front end of the "bird." This bird represents the parental generation.

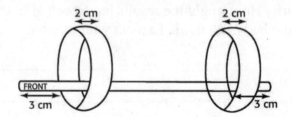

3. Test how far your parent bird can fly by releasing it with a gentle overhand pitch. Test the bird twice. Record the bird's average flight distance in a data table like the one shown.

Holt Biology Evolutionary Theory

Name _____ Class _____ Date _____

Natural Selection Simulation *continued*

Bird	Coin flip (H or T)	Die throw (1–6)	Anterior wing (cm)			Posterior wing (cm)			Avg. Dist. flown (m)
			Width	Circ.	Dist. from front	Width	Circ.	Dist. from back	
Parent	NA	NA	2	19	3	2	19	3	
Generation 1									
Chick 1									
Chick 2									
Chick 3									
Generation 2									
Chick 1									
Chick 2									
Chick 3									
Generation 3									
Chick 1									
Chick 2									
Chick 3									
Generation 4									
Chick 1									
Chick 2									
Chick 3									

Original content Copyright © by Holt, Rinehart and Winston. Additions and changes to the original content are the responsibility of the instructor.

Name _____ Class _____ Date _____
Natural Selection Simulation *continued*

Bird	Coin flip (H or T)	Die throw (1–6)	Anterior wing (cm)			Posterior wing (cm)			Avg.
			Width	Circ.	Dist. from front	Width	Circ.	Dist. from back	Dist. flowr (m)
Parent	NA	NA	2	19	3	2	19	3	
Generation 5									
Chick 1									
Chick 2									
Chick 3									
Generation 6									
Chick 1									
Chick 2									
Chick 3									
Generation 7									
Chick 1									
Chick 2									
Chick 3									
Generation 8									
Chick 1									
Chick 2									
Chick 3									

Original content Copyright © by Holt, Rinehart and Winston. Additions and changes to the original content are the responsibility of the instructor.

Holt Biology — Evolutionary Theory

MODEL FIRST (F₁) GENERATION

4. Each origami bird lays a clutch of three eggs. Assume that one of the chicks is identical to the parent. Use the parent data to fill in your data table for the first new chick (Chick 1).

5. Make two more chicks (Chick 2 and Chick 3). Assume that these chicks have mutations. Follow steps A through C for each chick to determine the effects of each mutation.

Step A Flip a coin to find out which end is affected by the mutation.

Heads = Front wing is affected.

Tails = Back wing is affected.

Step B Throw a die to find out how the mutation affects the wing.

- ⚀ = Wing position moves 1 cm toward the end of the straw.
- ⚁ = Wing position moves 1 cm toward the middle.
- ⚂ = Wing circumference increases by 2 cm.
- ⚃ = Wing circumference decreases by 2 cm.
- ⚄ = Wing width increases by 1 cm.
- ⚅ = Wing width decreases by 1 cm.

Step C If a mutation causes a wing to fall off the straw or makes a wing's circumference smaller than the circumference of the straw, the chick cannot "survive." If such a mutation occurs, record it as "lethal," and then produce another chick.

6. For each new chick, record the mutation and the new dimensions of each wing.

7. Test each bird twice by releasing it with a gentle overhand pitch. Release the bird as uniformly as possible. Record the distance that each bird flies. The most successful bird is the one that flies the farthest.

MODEL SUBSEQUENT GENERATIONS

8. Assume that the most successful bird in the previous generation is the sole parent of the next generation. Using this bird, repeat steps 4–7.

9. Continue to produce chicks and to test and record data for eight more generations.

Name _____ Class _____ Date _____

Natural Selection Simulation *continued*

CLEAN UP AND DISPOSE

10. Clean up your work area and all lab equipment. Return lab equipment to its proper place. Dispose of paper scraps in the designated waste container. Wash your hand thoroughly before you leave the lab and after you finish all work.

Analyze and Conclude

1. **Summarizing Results** Describe any patterns in the evolution of the birds in your model.

   _____
   _____
   _____
   _____

2. **Evaluating Models** How well does this lab model natural biological processes? What are the limitations of this model?

   _____
   _____
   _____
   _____

3. **Analyzing Data** Compare your data with your classmates' data. Identify any similarities and differences. Try to explain any trends that you notice in terms of the theory of natural selection.

   _____
   _____
   _____
   _____
   _____
   _____

Name _____ Class _____ Date _____

Natural Selection Simulation *continued*

Extensions

4. **Design an Experiment** Propose a new hypothesis about natural selection that you could test by observing real organisms. Write a brief proposal describing an experiment that could test this hypothesis. Be sure to give your prediction, explain your methods, identify variables, and plan for control groups.

Name _____ Class _____ Date _____

Inquiry Practice

DATASHEET B FOR IN-TEXT LAB

Genetic Drift

OBJECTIVES

- **Investigate** the effect of population size on genetic drift.
- **Analyze** the mathematics of the Hardy-Weinberg principle.

MATERIALS

- buttons, blue (10 to 100)
- buttons, white (10 to 100)
- buttons, red (10 to 100)
- jar or beaker, large, plastic

Preparation

1. **Scientific Methods State the Problem** How does population size affect allele frequencies? Read the procedure to see how you will test this.
2. **Scientific Methods Form a Hypothesis** For a hypothesis that predicts the results of this procedure for three different population sizes.

_____

_____

Procedure

1. Prepare to model the populations. First, assign each color button to one of the alleles (I^A, I^B, or i) of the ABO blood types. Notice how each possible pairing of alleles matches one of the four types (A, B, AB, or O). Then choose three different population sizes. Also choose one ratio of alleles at which to start all three populations (for example, I^A: I^B: i = 2: 2: 1). Create tables for your data.

2. Represent the first population's alleles by placing the appropriate number of blue, red, and white buttons in a jar.

3. Randomly select two buttons from the jar to represent one person. Record this person's genotype and phenotype. Tally the total number of each allele within this generation.

4. Repeat step 3 until you have modeled the appropriate number of people in the population. Place the buttons back into the jar.

5. Empty the jar. Refill it with the number and color of buttons that matches the tallies recorded in step 4.

6. Repeat steps 3 through 5 until you have modeled four generations.

Name _____ Class _____ Date _____
Genetic Drift *continued*

7. Repeat steps 2 through 6 to model two more populations.

Analyze and Conclude

1. **Analyzing Data** Describe any changes in genotype and phenotype ratios within each population over time.

   _____

   _____

   _____

   _____

2. **Explaining Results** Did any population maintain genetic equilibrium? Explain how you can tell.

   _____

   _____

   _____

3. **Scientific Methods** **Analyzing Results** Which population showed the greatest amount of genetic drift? Explain.

   _____

   _____

   _____

   _____

4. **Applying Concepts** How might your results differ if you started with a larger number of buttons? a smaller number? a different ratio of colors? Explain your answers.

   _____

   _____

   _____

   _____

   _____

   _____

Name _____ Class _____ Date _____

Genetic Drift *continued*

5. **Research** Use library and Internet resources to find out about the *genetic bottleneck* that seems to have occurred among cheetah populations in the past. Explain the meaning of this term and its consequences for cheetahs as a species.

_____

_____

_____

_____

_____

_____

Name _____ Class _____ Date _____

Skills Practice

DATASHEET B FOR IN-TEXT LAB

Dichotomous Keys

OBJECTIVES

- **Identify** objects by using a dichotomous key.
- **Design** a dichotomous key for a group of objects.

MATERIALS

- objects, common (6 to 10)
- labels, adhesive
- pencil

Procedure

USE A DICHOTOMOUS KEY

1. Work with a small group. Use the dichotomous key below to identify the leaves shown on p. 438 of your textbook. Identify one leaf at a time. Always start with the first 2 statements (1a and 1b). Follow the direction beside the statement that describes the leaf.

2. Proceed through the key until you get to the name of a tree. Record your answer for each leaf shown on the following sheet.

	Key to Forest Trees	
1a	Leaf edge is smooth or barely curved.	go to 2
1b	Leaf edge has teeth, waves, or lobes.	go to 3
2a	Leaf has a sharp bristle at its tip.	shingle oak
2b	Leaf has no bristle at its tip.	go to 4
3a	Leaf edge has small, shallow teeth.	Lombardy poplar
3b	Leaf edge has deep waves or lobes.	go to 5
4a	Leaf is heart shaped.	eastern redbud
4b	Leaf is not heart shaped	live oak
5a	Leaf edge has more than 20 large lobes.	English oak
5b	Leaf edge has more than 20 waves.	chestnut oak

Original content Copyright © by Holt, Rinehart and Winston. Additions and changes to the original content are the responsibility of the instructor.

Holt Biology — Classification

Name _____ Class _____ Date _____

Dichotomous Keys *continued*

A	
B	
C	
D	
E	
F	

DESIGN A DICHOTOMOUS KEY

3.  Put on safety goggles, gloves, and a lab apron. Chose 6 to 10 objects from around the classroom or from a collection supplied by your teacher. Before you go to the next step, have your teacher approve the objects your group has chosen.

4. Study the structure and organization of the dichotomous key, which includes pairs of contrasting descriptions that form a "tree" of possibilities. Use this key as a model for the next step.

5. Work with the members of your group to design a new dichotomous key for the objects that your group selected. Be sure that each part of the key leads to either a definite identification of an object or another set of possibilities. Be sure that every object is included.

6. Test your key using each one of the objects in your collection.

EXCHANGE AND TEST KEYS

7. After each group has completed the steps above, exchange your key and your collection of objects with another group. Use the key you receive to identify each of the new objects. If the new key does not work, return it to the group so corrections can be made.

CLEANUP

8.  Clean up your work area and return or dispose of materials as directed by your teacher. Wash your hands thoroughly before you leave the lab and after you finish all of your work.

Name _____ Class _____ Date _____
Dichotomous Keys *continued*

Analyze and Conclude

1. **Summarizing Data** List the identity of the tree for each of the leaves you analyzed in step 2.

   _____
   _____
   _____
   _____

2. **Scientific Methods Critiquing Procedures** What other characteristics might be used to identify leaves by using a dichotomous key?

   _____
   _____

3. **Analyzing Results** What challenges did your group face while making your dichotomous key?

   _____
   _____
   _____
   _____

4. **Evaluating Results** Were you able to use another group's key to identify the group's collection of objects? Describe your experience.

   _____
   _____
   _____
   _____

5. **Scientific Methods Analyzing Methods** Does a dichotomous key begin with general descriptions and then proceed to more specific descriptions, or vice versa? Explain your answer by using examples.

   _____
   _____
   _____
   _____

Name _____ Class _____ Date _____

Dichotomous Keys *continued*

6. **Scientific Methods Evaluating Methods** Is a dichotomous key the same as the Linnaean classification system? Explain your answer.

_____
_____
_____
_____

Extension

7. **Research** Do research in the library or media center to find out what types of methods, other than dichotomous keys, are used to identify organisms.

_____
_____
_____
_____
_____
_____

Name _____ Class _____ Date _____

Skills Practice

DATASHEET B FOR IN-TEXT LAB

Model of Rock Strata

OBJECTIVES

- **Model** the formation and analysis of strata.
- **Apply** the criteria used to identify index fossils on the strata model.
- **Evaluate** the effectiveness of the model to illustrate relative fossil age.

MATERIALS

- graduated cylinder, 100 mL
- aquarium gravel, four distinct colors
- tablespoon
- water, tap
- dish, small (8 per group)
- beans, dried (navy, black, pinto)

Procedure

1. Work in groups of three or four. Each student in the group should make a separate model. You will build up a series of layers in a column. You will model eight periods of time using different colors of gravel and different beans. The gravel represents sediment and the beans represent fossils. One tablespoon represents deposition that occurs over a 10,000-year period.

2. **CAUTION: Glass items such as graduated cylinders are fragile and may break.** Add 30 mL of tap water to the graduated cylinder.

3. For the first time period, choose a color. Have each member of the group add 1 Tbsp of that gravel color to their column. Randomly choose one member of the group to omit this layer.

4. Repeat step 3 using another color of your choice until you have modeled eight time periods. At the third time period, insert some navy beans; at the fifth time period, insert pinto beans; at the seventh time period, insert black beans. Record the strata order used by your group. Keep this record as a key to your models.

_____
_____
_____

Name _____ Class _____ Date _____

Model of Rock Strata *continued*

5. Exchange the models from your group with those of another group. Try to determine the order of strata used by that group.

6. Clean up your lab materials according to your teacher's instructions. Wash your hands before leaving the lab.

Analyze and Conclude

1. **Recognizing Relationships** Explain how this model relates to how sedimentary strata are formed.

   _____
   _____
   _____
   _____

2. **Scientific Methods Analyzing Conclusions** Describe your success at inferring the other group's strata order.

   _____
   _____

3. **Scientific Methods Inferring Conclusions** Compare the occurrence of the three types of "fossils" across the models from each group. What is the significance of these fossils? Explain your reasoning.

   _____
   _____
   _____
   _____

4. **Analyzing Models** If the same fossils are contained within different kinds of strata, can they be classified as index fossils? Explain.

   _____
   _____
   _____
   _____

Name _____ Class _____ Date _____

Skills Practice Lab   DATASHEET B FOR IN-TEXT LAB

Bacterial Staining

OBJECTIVES

- **Prepare** and **stain** smears of bacteria.
- **Practice** using sterile technique to avoid contaminating bacterial cultures.

MATERIALS

- paper towels
- microscope slides (3)
- culture tubes of bacteria (3)
- sterile cotton swabs
- beaker, 150 mL
- methylene blue stain in dropper bottle
- 70% isopropyl alcohol
- pencil, wax
- Bunsen burner with striker
- test-tube rack
- forceps or wooden alligator-type clothespin
- water, 75 mL
- microscope, compound

Procedure

1. Put on safety goggles, gloves, and a lab apron.

2. **CAUTION: Alcohol is flammable. Do not use alcohol in the room when others are using a Bunsen Burner.** Use alcohol and paper towels to clean the surface of your lab table and gloves. Allow the table to air-dry.

3. **CAUTION: Microscope slides are fragile and have sharp edges.** Use a wax pencil to label three microscope slides "A," "B," and "C."

4. **CAUTION: Keep combustibles such as alcohol-soaked paper towels away from flames. Do not light a Bunsen burner when others in the room are using alcohol.** Have your teacher light a Bunsen burner with a striker.

Bacterial Staining *continued*

PART A: MAKING A SMEAR

5. Remove the cap from culture tube A. **CAUTION: Keep the cap in your hand.** To avoid contaminating your bacterial culture, do not place the cap on the table or other surface.

6. Pass the opening of the tube through the flame of a Bunsen burner to sterilize the end of the culture tube.

7. Use a sterile swab to collect a small sample of bacteria by lightly touching the tip of the swab to the bacterial culture.

8. Pass the opening of the tube through the flame again, and replace the cap.

9. Make a smear of bacterial culture A by rubbing the swab on the slide. Spread a thin layer of culture over the middle area of the slide. Cover about half of the total slide area and allow to dry.

10. Dispose of the swab in a proper container.

11. Repeat steps 5 through 10 for cultures B and C.

PART B: STAINING BACTERIA

12. Using microscope slide forceps, pick up each slide one at a time, and pass it over the flame several times. Let each slide cool.

13. Using microscope slide forceps, place one of your slides across the mouth of a 150 mL beaker half-filled with water.

14. **CAUTION: Methylene blue will stain your skin and clothing.** Place 2 to 3 drops of methylene blue stain on the dried bacteria. Do not allow the stain to spill into the beaker.

15. Let the stain stay on the slide for 2 min.

16. Dip the slide into the water in the beaker several times to rinse it. Gently blot the slide dry with a paper towel. Do not rub the slide.

17. Repeat steps 14 through 17 for your other two slides.

18. Allow each slide to completely dry before observing your slides under the microscope.

Name _____ Class _____ Date _____

Bacterial Staining *continued*

PART C: OBSERVING BACTERIA

19. Observe each slide under the microscope on low and high power. Make a sketch of a few cells that you see on each slide.

Slide A

Slide B

Slide C

Name _____ Class _____ Date _____

Bacterial Staining *continued*

20. Clean up your lab materials according to your teacher's instructions. Wash your hands before leaving the lab.

Analyze and Conclude

1. **Summarizing Results** Describe the shape and grouping of the cells of each type of bacteria that you observed.

   _____
   _____
   _____

2. **Drawing Conclusions** How did you classify the bacteria in cultures A, B, and C: as coccus, bacillus, or spirillum?

   _____
   _____
   _____

3. **Evaluating Viewpoints** Evaluate the following advice: Always use caution when handling bacteria, even if the bacteria are known to be harmless.

   _____
   _____
   _____
   _____
   _____

Name _____ Class _____ Date _____
Bacterial Staining *continued*

Extensions

4. **Further Inquiry** Write a question about bacteria that could be explored with another investigation using skills that you learned in this lab.

_____
_____
_____

5. **On the Job** Microbiologists are scientists who study organisms too small to be seen by the unaided eye. Do research to find out about the kinds of work that microbiologists do and how microbiologists improve our lives.

_____
_____
_____
_____
_____
_____
_____
_____
_____

Name _____ Class _____ Date _____

Inquiry **DATASHEET B FOR IN-TEXT LAB**

Protistan Responses to Light

OBJECTIVES

- **Identify** several types of protists.
- **Compare** the structures, methods of locomotion, and behaviors of several kinds of protists.
- **Relate** a protist's response to light to the protist's method of feeding.

MATERIALS

- protist slowing agent
- mixed culture of protists
- microscope slides
- toothpicks
- paper, white
- scissors
- sunlit windowsill or lamp
- plastic pipets with bulbs
- compound microscope
- coverslips
- construction paper, black
- paper punch
- forceps

Preparation

1. **Scientific Methods** **State the Problem** How do protists respond to various amounts of light?
2. **Scientific Methods** **Form a Hypothesis** Form a testable hypothesis about how protists will respond to various levels of light.

_____

_____

Procedure

MAKE A WET MOUNT OF PROTISTS

1.  Put on safety goggles, gloves, and a lab apron.

Holt Biology — Protists

Name _____ Class _____ Date _____

Protistan Responses to Light *continued*

2. **CAUTION: Do not touch your face while handling microorganisms.** Place a drop of protist slowing agent on a microscope slide. Add a drop of liquid from the bottom of a mixed culture of protists. Add a coverslip.

3. View the slide under low power and high power of a microscope.

4. Make a drawing of each type of protist. Note whether the protist moves, and try to determine how it moves.

Movement:	Movement:
Movement:	Movement:

Holt Biology — Protists

Protistan Responses to Light *continued*

5. Repeat step 1, but do not use slowing agent. Note differences in the movement of the protists that you see.

_____

_____

TEST PROTISTAN RESPONSES TO LIGHT

6. Place a wet mount of protists on a piece of white paper. Then, place the paper and the slide on a sunlit windowsill or under a table lamp.

7. Punch a hole in a piece of black construction paper that has a slight curl, as shown in the photo on p. 513 of your textbook. Position the black paper on top of the slide so that the hole is in the center of the coverslip.

8. To examine the slide, first view the area in the center of the hole under low power. (Note: Do not disturb the black paper, and do not switch to high power. Switching to high power will move the paper.) Then, have a partner carefully remove the black paper with forceps while you observe the slide. Note any movement of the protists in response to the change in light.

_____

_____

_____

_____

DESIGN AN EXPERIMENT

9. Design an experiment that tests your hypothesis and that uses the materials listed for this lab. Predict what will happen during your experiment if your hypothesis is supported.

_____

_____

_____

_____

Name _____ Class _____ Date _____
Protistan Responses to Light *continued*

10. Write a procedure for your experiment. Identify the variables that you will control, the experimental variables, and the responding variables. Construct any tables you will need to record your data. Make a list of all safety precautions you will take. Have your teacher approve your procedure before you begin.

_____
_____
_____
_____
_____
_____
_____
_____

11. Set up and carry out your experiment.
12. Clean up your lab materials according to your teacher's instruction. Wash your hands before you leave the lab.

Analyze and Conclude

1. **Summarizing Results** Describe the various types of locomotion that you observed in protists, and give examples of each type.

_____
_____
_____
_____

2. **Scientific Methods Analyzing Results** Identify which protists were affected by light, and describe how they were affected.

_____
_____
_____
_____
_____

Protistan Responses to Light continued

3. **Scientific Methods Drawing Conclusions** How are a protist's response to light and the protist's method of feeding related?

___

Extensions

4. **Research** Investigate livestock diseases that are caused by parasitic protists. Which of these diseases are most common in the United States?

___

5. **Research** Find out how backpackers can avoid getting diseases that are caused by protists and transmitted in water.

___

Name _____ Class _____ Date _____

Inquiry

DATASHEET B FOR IN-TEXT LAB

Yeast and Fermentation

OBJECTIVES

- **Observe** the process of fermentation by yeast.
- **Investigate** various energy sources and their effects on fermentation.
- **Measure** energy released by fermentation in the form of heat.

MATERIALS

- vacuum flask, 500 mL
- beaker, 250 mL
- glucose (75 g)
- potato flakes (50 mL)
- packets of artificial sweetener (5)
- thermometer
- rubber stopper, one-hole
- sucrose (75 g)
- milk (50 mL)
- package of dry baker's yeast
- rubber tubing, 50 cm long

Preparation

1. **Scientific Methods State the Problem** Which carbohydrate does yeast use most efficiently as food for fermentation?
2. **Scientific Methods Form a Hypothesis** Form a testable hypothesis about which carbohydrate will promote yeast fermentation the best.

_____

_____

Procedure

1.  Put on safety goggles, gloves, and a lab apron.

2. Mix 75 g or 50 mL of a carbohydrate of your choice in 400 mL of water.

3. **CAUTION: Do not touch your face while working with active yeast.** When your carbohydrate is thoroughly mixed, add one package of dry yeast and stir. Pour the yeast solution into a vacuum flask.

Name _____ Class _____ Date _____
Yeast and Fermentation *continued*

4. **CAUTION: Use extreme caution when working with glass.** Put a thermometer in a one-hole rubber stopper. Put the stopper in the mouth of the open bottle. Adjust the thermometer so that it sticks in the yeast solution.

5. Attach a piece of rubber tubing to the side arm of your vacuum flask. Place the end of the tubing into a beaker of water so that you can observe bubbles of CO_2 produced during fermentation.

6. Record the temperature of the solution. Continue to record the temperature at regular intervals during the next two days.

Time	Temperature

9.  Clean up your lab materials according to your teacher's instruction. Wash your hands before you leave the lab.

Analyze and Conclude

1. **Summarizing Data** Prepare a line graph of your data to illustrate the temperature of your solution over time.

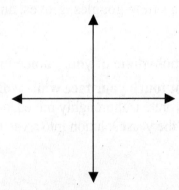

Name _____ Class _____ Date _____

Yeast and Fermentation *continued*

2. **Scientific Methods** **Analyzing Data** Compare your graph to your classmates' graphs. Which carbohydrate best supported yeast growth.

_____

_____

3. **Scientific Methods** **Analyzing Methods** Design an experiment in which you measure CO_2 production in a different way.

_____

_____

_____

_____

_____

_____

Name _____ Class _____ Date _____

Skills Practice Lab

DATASHEET B FOR IN-TEXT LAB

Plant Diversity

OBJECTIVES

- **Identify** similarities and differences among four phyla of living plants
- **Relate** structural adaptations of plants to plants' success on land.

MATERIALS

- live or preserved specimens of mosses, ferns, conifers, and
- prepared slides of fern gametophytes angiosperms
- stereomicroscope or hand lens
- prepared slides of pine pollen
- compound microscope

Procedure

1. Visit the station for each of the plants listed below, and examine the specimens there. Record your observations.

Name _____ Class _____ Date _____

Plant Diversity *continued*

2. **Mosses** Use a stereomicroscope or hand lens to examine a moss gametophyte that has a sporophyte attached to it. Draw what you see, and label the parts that you recognize.

3. **Ferns** Examine the sporophyte of a fern, and look for evidence of reproductive structures on the underside of the fronds. Use a compound microscope to examine a slide of a fern gametophyte. Draw what you see, and label any structures you recognize.

Name _____ Class _____ Date _____

Plant Diversity *continued*

4. **Conifers** Draw a part of a branch of one of the conifers at this station. Label a leaf, stem, and cone (if present). Examine a prepared slide of pine pollen. Draw a few of the pollen grains.

5. **Angiosperms** Draw one of the representative angiosperms at this station. Label a leaf, stem, root, and flower (if present). Indicate the sporophyte and location of gametophytes.

6. Clean up your lab materials according to your teacher's instructions. Wash your hands before leaving the lab.

Name _____ Class _____ Date _____

Plant Diversity *continued*

Analyze and Conclude

1. **Recognizing Patterns** How do the gametophytes of gymnosperms and angiosperms differ from the gametophytes of mosses and ferns?

   _____
   _____
   _____
   _____

2. **Comparing Structures** What structures are present in both gymnosperms and angiosperms but absent in both mosses and ferns?

   _____
   _____
   _____
   _____
   _____

Name _____ Class _____ Date _____

Skills Practice Lab

DATASHEET B FOR IN-TEXT LAB

Monocot and Dicot Seeds

OBJECTIVES

- **Observe** the structures of bean seeds and corn kernels.
- **Compare** and **contrast** the development of bean embryos as they grow into seedlings.

MATERIALS

- bean seeds, soaked overnight (6)
- corn kernels, soaked overnight (6)
- paper towels
- beakers, 150 mL (2)
- ruler, metric
- stereomicroscope
- scalpel
- rubber bands (2)
- pen, glass-marking

Procedure

OBSERVE SEED STRUCTURE

1. Remove the seed coat of a bean seed, and separate the two fleshy halves of the seed.

2. Locate the embryo on one of the halves of the seed. Examine the bean embryo with a stereomicroscope. Draw the embryo in the box below, and label the parts that you can identify.

3. 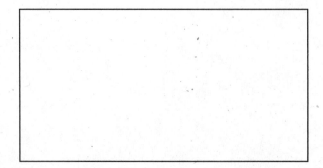 **CAUTION: Put on goggles before you handle scalpels or glassware. Sharp or pointed objects may cause injury.** Handle scalpels carefully. Examine a corn kernel and locate a small, light-colored oval area. Use a scalpel to cut the kernel in half along the length of this area

Name _____ Class _____ Date _____

Monocot and Dicot Seeds *continued*

4. Locate the corn embryo, and examine it with a stereomicroscope. Draw the embryo in the box below, and label the parts that you can identify.

OBSERVE SEEDLING DEVELOPMENT

5. Fold a paper towel in half. Set five corn kernels on the paper towel.
6. Roll up the paper towel. Put a rubber band around the roll.
7. **CAUTION: Use glass beakers carefully.** Stand the roll in a beaker with 1 cm of water in the bottom.
8. Add water to the beaker as needed to keep the paper towels wet, but do not allow the corn kernels to be covered by water.
9. Repeat step 5 with five bean seeds.
10. After three days, unroll the paper towels, and examine the corn and bean seedlings.
11. Use the glass-marking pen to mark the roots and shoots of the developing seedlings. Starting at the seed, make a mark every 0.5 cm along the root of each seedling. Again, starting at the seed, make a mark every 0.5 cm along the stem of each seedling.
12. Draw a corn seedling and a bean seedling below. Label the parts of each seedling. Also, show the marks you made on each seedling, and indicate the distance between the marks.

Corn seedling	Bean seedling

13. Roll up the seeds in a fresh paper towel. Place the rolls back in the beakers and add fresh water.
14. After two more days, look at the seedlings. Measure the distance between the marks. Repeat step 8.

Name _____ Class _____ Date _____

Monocot and Dicot Seeds *continued*

15. Clean up your lab materials according to your teacher's instructions. Wash your hands before leaving the lab.

Analyze and Conclude

1. **Relating Concepts** Corn and beans are often cited as representative examples of monocots and dicots, respectively. Relate the seed structure of each to the terms *monocotyledon* and *dicotyledon*.

   _____
   _____
   _____
   _____

2. **Scientific Methods Summarizing Results** What parts of a plant embryo were observed in all seedlings on the third day?

   _____
   _____

3. **Drawing Conclusions** In which part or parts of bean seedlings and corn seedlings do the seedlings grow in length? Explain.

   _____
   _____
   _____

4. **Scientific Methods Forming Hypotheses** How are the tender young shoots of bean seedlings and corn seedlings protected as the seedlings grow through the soil?

   _____
   _____
   _____

5. **Evaluating Viewpoints** Defend the following statement: There are both similarities and differences in seed structure and seedling development in beans and corn.

   _____
   _____
   _____
   _____

Name _____ Class _____ Date _____
Monocot and Dicot Seeds *continued*

Extensions

6. **Further Inquiry** Write a new question about seedling development that could be explored with another investigation.

   _____

   _____

   _____

7. **Career Connection** Plant physiology is the study of the processes that occur in plants. Do research to discover where plant physiologists work and what types of research are currently being conducted in the field of plant physiology.

Name _____ Class _____ Date _____

Skills Practice   DATASHEET B FOR IN-TEXT LAB

Cultivation Techniques

OBJECTIVES

- **Compare** hydroponic plant-cultivation with conventional plant-cultivation techniques.
- **Observe** the germination of wheat seeds over a two-week period.

MATERIALS

- pen, marking
- clear plastic cups (2)
- wheat seeds (12)
- water, distilled, 10 mL
- complete nutrient solution, 50 mL
- dropper, plastic
- ruler, metric
- tape, labeling
- potting soil, 50 mL
- plastic-foam floater that has 6 evenly spaced holes
- graduated cylinder, 50 mL
- cheesecloth, large enough to cover the plastic-foam floater

Procedure

DAY 1

1.  Put on a lab apron, safety goggles, and disposable gloves.

2. Using the marking pen and the labeling tape, label one plastic cup "Soil cultivated," and label the other plastic cup "Hydroponically cultivated."

3. Fill the cup labeled "Soil cultivated" halfway with moist potting soil. Place six wheat seeds on the surface of the soil; use the distance between the holes in the foam floater as a guide to determine the spacing of the wheat seeds. (Do not place the floater on the soil.)

4. Press the seeds into the soil until they are approximately 0.5 cm below the surface. Cover the seeds with soil, and press down firmly.

5. Water the seeds with 10 mL of distilled water.

6. Add 50 mL of complete nutrient solution to the cup labeled "Hydroponically cultivated," and place the plastic-foam floater on the surface of the solution.

Name _____ Class _____ Date _____

Cultivation Techniques *continued*

7. Place the cheesecloth on top of the floater. Press lightly at the location of the holes in the floater to moisten the cheesecloth.

8. Place the remaining six wheat seeds on top of the cheesecloth in the cup labeled "Hydroponically cultivated." Position the seeds so that each one lies in an indentation formed by the cheesecloth in a hole in the floater. Press each seed lightly into the hole until the seed coat is moistened.

9. Place both cups in a warm, dry location. Water the soil-cultivated seeds as needed, and monitor the amount of water added. Aerate the roots of the hydroponic plants every day by using a clean plastic dropper to blow air into the nutrient solution.

10. Use the data tables below to write your observations of the seeds.

11. Clean up your lab materials according to your teacher's instructions. Wash your hands before leaving the lab.

Observations of Soil-Grown Plants

Day	Appearance of seedlings	Average height (mm)
1		
2		
3		
4		
5		
6		
7		
8		
9		
10		
11		
12		
13		
14		

Cultivation Techniques *continued*

Observations of Hydroponically Grown Plants

Day	Appearance of seedlings	Average height (mm)
1		
2		
3		
4		
5		
6		
7		
8		
9		
10		
11		
12		
13		
14		

DAYS 2–14

12. Compare the contents of each cup every day for two weeks, and record the appearance of the wheat seedlings in your data tables. If you are unable to observe your seedlings over the weekend, be sure to note in your data tables that no observations were made on those days.

13. Each time that you observe the seedlings after they have begun to grow, measure their height and record in your data tables the average height of the seedlings in each cup. To find the average height for one cup, add the heights of each seedling in the cup together and divide by the number of seedlings.

14. After the seeds in the cup containing nutrient solution have germinated and formed roots, allow an air pocket to form between the floater and the surface of the nutrient solution. A portion of the roots should still be submerged in the nutrient solution. The air pocket allows the roots of the seeds to absorb the oxygen necessary for metabolic processes while continuing to absorb nutrients from the nutrient solution. Continue to observe and record the progress of the seedlings in each cup on a daily basis.

15. Clean up your lab materials according to your teacher's instructions. Wash your hands before leaving the lab.

Name _____ Class _____ Date _____
Cultivation Techniques *continued*

Analyze and Conclude

1. **Analyzing Data** Based on the data that you recorded, which seeds germinated more quickly? Which seeds grew taller?

   _____
   _____

2. **Scientific Methods Analyzing Results** Compare your results with those of your classmates. Were the results the same for each group?

   _____
   _____
   _____

3. **Scientific Methods Analyzing Methods** Why do you think that you were instructed to plant six seeds instead of a single seed in each cup? Why is the use of more than one sample important?

   _____
   _____
   _____

Extensions

4. **Further Inquiry** The nutrient solution that you used in this investigation should have provided all of the inorganic nutrients that the wheat seeds needed for proper growth. How could you determine, by using hydroponic cultivation, exactly which organic nutrients a plant requires?

   _____
   _____
   _____
   _____
   _____

Name _____ Class _____ Date _____

Skills Practice

DATASHEET B FOR IN-TEXT LAB

Embryonic Development

OBJECTIVES

• **Identify** the stages of early animal development.
• **Describe** the changes that occur during early development.

MATERIALS

- prepared slides of sea-star development, including:
  unfertilized egg
  zygote
  2-cell stage
  4-cell stage
  8-cell stage
  16-cell stage
  32-cell stage
  64-cell stage
  blastula
  early gastrula
  middle gastrula
  late gastrula
  young sea-star larva
- compound light microscope
- paper and pencil

Procedure

1. Obtain a set of prepared slides that show sea-star eggs at different stages of development. Choose slides labeled unfertilized egg, zygote, 2-cell stage, 4-cell stage, 8-cell stage, 16-cell stage, 32-cell stage, 64-cell stage, blastula, early gastrula, middle gastrula, late gastrula, and young sea-star larva. (Note: Blastula is the general term for the embryonic stage that results from cleavage. In mammals, a blastocyst is a modified form of the blastula.)

Name _____ Class _____ Date _____
Embryonic Development *continued*

2. **CAUTION: Glassware is fragile.** Examine each slide using a compound light microscope. Using the microscope's low-power objective first, focus on one good example of the developmental stage listed on the slide's label. Then switch to the high-power objective, and focus on the image with the fine adjustment. Notify the teacher of broken glass or cuts. Do not clean up broken glass or spills with broken glass unless the teacher tells you to do so.

3. Below, draw a diagram of each developmental stage that you examine (in chronological order). Label each diagram with the name of the stage it represents and the magnification used. Record your observations as soon as they are made. Do not redraw your diagrams. Draw only what you see; lab drawings do not need to be artistic or elaborate. They should be well organized and include specific details.

4. Clean up your lab materials according to your teacher's instruction. Wash your hands before you leave the lab.

Developmental Stages

Name _____ Class _____ Date _____

Embryonic Development *continued*

Developmental Stages

Name _____ Class _____ Date _____
Embryonic Development *continued*

Analysis and Conclusions

1. **Summarizing Results** Compare the size of the sea-star zygote with that of the blastula. At what stage does the embryo become larger than the zygote?

   _____

2. **Analyzing Data** At what stage do all of the cells in the embryo not look exactly like each other? How do cell shape and size change during successive stages of development?

   _____
   _____
   _____

3. **Drawing Conclusions** From your observations of changes in cellular organization, why do you think the blastocoel (the space in the center of the hollow sphere of cells of a blastula) is important during embryonic development?.

   _____
   _____
   _____
   _____

4. **Predicting Patterns** How are the symmetries of a sea-star embryo and a sea-star larva different from the symmetry of an adult sea star? Would you expect to see a similar change in human development? What must happen to the sea-star gastrula before it becomes a mature sea star?

   _____
   _____
   _____
   _____
   _____

Name _____ Class _____ Date _____
Embryonic Development *continued*

Extensions

5. **Further Inquiry** Using the procedure that you followed in this investigation, compare embryonic development in other organisms with embryonic development in sea stars. Which types of organisms would you expect to develop similarly to sea stars? Which types of organisms would you expect to develop differently from sea stars?

_____

_____

_____

_____

_____

_____

_____

_____

_____

Name _____ Class _____ Date _____

Skills Practice Lab   DATASHEET B FOR IN-TEXT LAB

Hydra Behavior

OBJECTIVES

- **Observe** a hydra finding and capturing prey.
- **Determine** how a hydra responds to stimuli.

MATERIALS

- microscope slide
- *hydra* culture
- filter paper, cut into pennant shapes
- beef broth, concentrated
- silicone culture gum
- eyedroppers (2)
- stereomicroscope
- forceps
- *Daphnia* culture

Procedure

1.  **CAUTION: Handle glass slides carefully. Handle live organisms with care.** Arrange a piece of silicon gum to form a circular well on a microscope slide, as shown in the photo on page 671 of your textbook.

2.  With an eyedropper, transfer a hydra from its culture dish to the well on the slide. Cover the animal in water. Examine the hydra under the microscope.

3. Hold a piece of filter paper with forceps and move the tip of the paper near, but not touching, the hydra's tentacles. Observe and record the hydra's response. Dip the same filter paper in beef broth and repeat the procedure. Record these results also.

4. Use the tip of a clean piece of filter paper to gently touch the hydra's tentacles, disk, and stalk. Record your observations.

5. Use the eyedropper to transfer live *Daphnia* to the well with the hydra on the microscope slide. Carefully observe the hydra under the microscope. Record your observations in a data table.

6. If the hydra does not respond, repeat steps 2–5 with another hydra.

7. Clean up your lab materials according to your teacher's instructions. Wash your hands before leaving the lab.

Name _____ Class _____ Date _____
Hydra Behavior *continued*

Analyze and Conclude

1. **Analyzing Results** Describe the hydra's response to touch, chemicals (beef broth), and prey *(Daphnia)*.

   _____
   _____
   _____
   _____

2. **Critical Thinking Drawing Conclusions** How does a hydra detect its prey? Give evidence to support your conclusions.

   _____
   _____

3. **Critical Thinking Making Predictions** Based on your observations, how do you think a hydra behaves when it detects a threat in its natural environment?

   _____
   _____
   _____
   _____
   _____

4. **Further Research** Find out what kinds of food hydras eat and how the feeding method of a hydra differs from that of a sponge.

   _____
   _____
   _____
   _____
   _____
   _____

Name _____ Class _____ Date _____

Skills Practice DATASHEET B FOR IN-TEXT LAB

Clam Characteristics

OBJECTIVES
- **Observe** the behavior of a live clam.
- **Examine** the structure and composition of a clamshell.

MATERIALS
- lab apron
- disposable gloves
- beaker or dish, small
- food coloring
- clamshell
- hammer, small
- HCl, 0.1 M
- safety goggles
- clam, live
- eyedropper
- stirring rod, glass
- Petri dish
- stereomicroscope

Procedure

OBSERVE A LIVE CLAM

1.  Put on a lab apron, safety goggles, and disposable gloves.

2.  **CAUTION: Glassware is fragile. Notify the teacher of broken glass or cuts. Do not clean up broken glass or spills that contain broken glass unless the teacher tells you to do so.** Place a live clam in a small beaker or shallow dish of water.

3. **CAUTION: Do not touch or taste any chemicals. Know the location of the emergency shower and eyewash station and how to use them. If you get a chemical on your skin or clothing, wash it off at the sink while calling to the teacher. Notify the teacher of a spill. Spills should be cleaned up promptly, according to your teacher's directions.** Using an eyedropper, apply two drops of food coloring near the clam.

4. Observe and record what happens to the food coloring.

_____

_____

Holt Biology 113 Clam Characteristics

Name _____ Class _____ Date _____
Clam Characteristics *continued*

5. **CAUTION: Touch the clam gently to avoid injuring it. Remember that the clam is a live animal.** Using a stirring rod, touch the clam's mantle.

6. Observe and record the clam's response to touch.

_____

_____

OBSERVE A CLAMSHELL

7. Examine the surface of the clamshell. Locate the knob-shaped umbo on the shell. If you have difficulty finding the umbo, use the diagram of a clam as a guide (p. 692 of your textbook). Take note of the concentric growth rings on the shell. Then, count and record the number of growth rings on the clamshell.

    _____ growth rings

8. Place the clamshell in a Petri dish.

9. Use a small hammer to chip away part of the shell in order to expose the three layers of the shell. View the shell's layers by using a stereomicroscope. The outermost layer protects the clam from acids in the water. The innermost layer is mother-of-pearl, the material that forms pearls.

10. **CAUTION: Hydrochloric acid is corrosive. Avoid contact with skin, eyes, and clothing. Avoid breathing vapors.** The middle layer of the shell contains crystals of calcium carbonate. To test for the presence of this compound, place one drop of 0.1 M HCl on the middle layer of the shell. If calcium carbonate is present, bubbles of carbon dioxide will form in the drop. Record your observations.

_____

_____

11. Clean up your lab materials according to your teacher's instructions. Wash your hands before leaving the lab.

Name _____ Class _____ Date _____
Clam Characteristics *continued*

Analyze and Conclude

1. **Scientific Methods Analyzing Data** Find the incurrent and excurrent siphons of the clam in the diagram (p. 692). Using this information, explain your observations in step 4.

   _____
   _____
   _____

2. **Drawing Conclusions** What is the purpose of a clam's shell?

   _____
   _____
   _____
   _____

3. **Scientific Methods Using Evidence to Develop Predictions** Based on your observations, how do you think clams respond when they are touched or threatened in their natural habitat?

   _____
   _____
   _____
   _____

4. **Scientific Methods Forming a Hypothesis** What does a clam take in from the water that passes through its body?

   _____
   _____

5. **Inferring Relationships** Water that enters a clam's incurrent siphon passes over the clam's gills. How does this help the clam respire?

   _____
   _____
   _____

Name _____ Class _____ Date _____
Clam Characteristics *continued*

Extensions

1. **Further Inquiry** Write a new question about clams that could be explored in another investigation.

   _____

   _____

   _____

2. **Designing an Investigation** Design an investigation that could answer the question that you wrote for the Further Inquiry extension exercise.

   _____

   _____

   _____

Name _____ Class _____ Date _____

Skills Practice Lab   DATASHEET B FOR IN-TEXT LAB

Butterfly Metamorphosis

OBJECTIVES

- **Create** and **maintain** a habitat for caterpillars and butterflies.
- **Observe** the stages of metamorphosis in a butterfly.
- **Observe** the feeding behaviors of caterpillars and butterflies.
- **Develop** hypotheses about insect food preferences.

MATERIALS

- disposable gloves
- butterflies, adult painted lady
- hand lens
- lettuce
- packing box, large
- sugar, white
- paper toweling
- tape
- caterpillars, painted lady (5)
- plate, small
- dandelion greens
- shoe box
- milk carton, small
- scissors
- mosquito netting

Procedure

OBSERVE FOOD PREFERENCES IN CATERPILLARS

1. Work in a small group. Put on disposable gloves.

2. **CAUTION: Exercise caution when using live animals.** Gently transfer a caterpillar that is at least 2 cm long to a small plate. Use a hand lens to observe this animal. Draw what you see.

Name _____ Class _____ Date _____

Butterfly Metamorphosis *continued*

3. Place several fresh dandelion greens on one side of a shoe box. Place a fresh lettuce leaf on the other side of the box.

4. Place four caterpillars in the box. Cover the box with mosquito netting, and tape into place. Place the box in a well-lit area, but not in direct sunlight. Maintain setup at room temperature.

5. Each day, record which food the caterpillars preferred in the table below. Then replace the wilted food with fresh greens and leaves.

Food Preference

Insect	Type of food	Number of individuals
Caterpillar	Dandelion leaf	
	Lettuce	

6. After about a week, you may observe that the caterpillars have secured themselves on the overhead netting and entered a new life stage. If so, let this hanging stage (called a *pupa*) remain undisturbed for 7 to 10 days. Go on to the next part of this activity.

MAINTAIN ADULT BUTTERFLIES

7. Carefully poke a hole about the diameter of a pencil into the top of a small and clean milk carton.

8. Open the carton spout, and fill it halfway with tap water. Add a packet of sugar. Close the spout, and swirl to mix the solution.

9. Twist a piece of paper towel into a tight roll. Insert this rolled towel through the carton hole so that it acts like a wick.

10. Undo the interwoven flaps forming the top and bottom of a large packing box. Use scissors to remove these eight cardboard flaps. Only a cardboard frame should remain.

11. Cut out two large rectangles of mosquito netting that can fit over each of the two openings. Stretch a piece of mosquito netting over one open side. Secure the fabric with tape.

Name _____ Class _____ Date _____

Butterfly Metamorphosis *continued*

12. Stand the box up on one of its sides. Set the milk carton feeder inside the frame on its base. Use the cardboard flaps to build structures on which the butterflies can perch inside the frame.
13. Begin securing the netting over the open side of the frame. Before completely sealing the chamber, introduce the adult butterflies.
14. Each day for five days, observe and record the butterfly feeding behavior. The butterflies should be exposed to sunlight for several hours each day.

Butterfly feeding pattern

Day 1	
Day 2	
Day 3	
Day 4	
Day 5	

OBSERVE FOOD PREFERENCES IN BUTTERFLIES

15. Assemble an additional feeder station. Change the identity of the liquid (use various sugar concentrations or sugar substitutes) or appearance of this second station (add petal like structures).
16. Offer both stations to the butterflies, and record the feeder around which the butterflies spend most of their time.

Food Preference

Insect	Type of food	Number of individuals
Butterfly	Sugar solution	
	Experimental solution	

Holt Biology   Butterfly Metamorphosis

Name _____ Class _____ Date _____

Butterfly Metamorphosis *continued*

17. Clean up your lab materials according to your teacher's instructions. Wash your hands before leaving the lab.

Analyze and Conclude

1. **Evaluating Results** Did the caterpillars demonstrate a food preference? If so, what did you observe?

   _____
   _____

2. **Recognizing Patterns** Compare feeding in caterpillars and adult butterflies. Record similarities and differences.

   _____
   _____
   _____
   _____

3. **Scientific Methods Identifying Variables** What was the experimental variable that you used to explore food preferences in butterflies? What did you learn?

   _____
   _____
   _____
   _____

4. **Scientific Methods Identifying Relationships Among Variables** Why might the diet of an adult butterfly contain a concentrated energy source while the food preferred by the caterpillar be less energy intensive?

   _____
   _____
   _____
   _____
   _____
   _____

Name _____ Class _____ Date _____

Butterfly Metamorphosis *continued*

Extensions

5. **Forming Reasoned Opinions** How might the petal of a flower offer an advantage to a feeding butterfly?

   _____
   _____
   _____
   _____

6. **Further Inquiry** Many arthropods and echinoderms have preferences for certain foods. Conduct research in the library or on the Internet, and make a table that describes food preferences in at least five types of arthropods or echinoderms. Try to formulate a hypothesis that addresses why each animal has a particular food preference.

   _____
   _____
   _____
   _____
   _____

Name _____ Class _____ Date _____

Skills Practice Lab

DATASHEET B FOR IN-TEXT LAB

Live Frog Observation

OBJECTIVES

- **Examine** the external features of a frog.
- **Observe** the behavior of a frog.
- **Explain** how a frog is adapted to life on land and in water.

MATERIALS

- lab apron, disposable gloves
- terrarium
- beaker, 600 mL
- dechlorinated water
- frog, live
- insects, live (crickets or mealworms)
- aquarium

Procedure

1. Observe a live frog in a terrarium. Closely examine the external features of the frog. Then make a drawing of the frog in the space below. Label the eyes, nostrils, tympanic membranes, front legs, and hind legs.

Name _____ Class _____ Date _____
Live Frog Observation *continued*

2. Use the data table below to keep track of your observations. Watch the frog's movements as it breathes air into and out of its lungs. Record your observations.

Behavior/structure	Observations
Breathing	
Eyes	
Legs	
Response to food	
Response to noise	
Skin	
Swimming behavior	

3. Look closely at the frog's eyes, and note their location. Examine the upper and lower eyelids as well as a third transparent eyelid called a nictitating membrane. Describe how the eyelids move.

4. Study the frog's legs, and note the difference between the front and hind legs.

5. Place a live insect, such as a cricket or a mealworm, into the terrarium. Observe and record how the frog reacts.

Name _____ Class _____ Date _____

Live Frog Observation *continued*

6. Gently tap the side of the terrarium farthest from the frog, and observe the frog's responses. Record your observations.

7. Put on gloves and a lab apron.

8. **CAUTION: Handle live frogs gently. Frogs are slippery! Do not allow a frog to injure itself by jumping from a lab table to the floor.** Place a 600 mL beaker in the terrarium. Carefully pick up the frog, and examine its skin. How does it feel? Now place the frog in the beaker. Cover the beaker with your hand and carry it to a freshwater aquarium. Tilt the beaker, and gently lower it into the water until the frog swims out.

9. Watch the frog float and swim. Notice how the frog uses its legs to swim. Also notice the position of the frog's head. As the frog swims, bend down to view the underside of the frog. Then look down on the frog from above. Compare the color on the dorsal and ventral sides of the frog.

10. Clean up your lab materials according to your teacher's instructions. Wash your hands before you leave the lab.

Analyze and Conclude

1. **Making Systematic Observations** How does a frog use its hind legs for moving on land and in water?

   _____
   _____

2. **Using Evidence to Develop Explanations** How does the position of a frog's eyes benefit the frog while it is swimming?

   _____
   _____
   _____
   _____

3. **Scientific Methods Using Evidence to Develop Predictions** What features of an adult frog provide evidence that it has an aquatic life and a terrestrial life?

   _____
   _____
   _____
   _____
   _____

Name _____ Class _____ Date _____

Live Frog Observation *continued*

4. **Scientific Methods Critiquing Explanations** Were you able to determine in this lab how a frog hears? Explain your reasoning.

_____
_____
_____
_____

Extensions

5. **Inferring Conclusions** What can you infer about a frog's field of vision from the position of its eyes?

_____
_____
_____
_____

6. **Forming Hypotheses** How is the coloration of the dorsal side of the frog an adaptive advantage?

_____
_____
_____

7. **Further Inquiry** Write a new question about frogs that could be explored with another investigation.

_____
_____
_____

8. **Careers in Science** Herpetology is the study of reptiles and amphibians. Do research to discover how herpetologists are working with the Declining Amphibian Task Force to solve the mystery of the worldwide decline in amphibian populations.

Name _____ Class _____ Date _____

Skills Practice Lab

DATASHEET B FOR IN-TEXT LAB

Bird Digestion

OBJECTIVES

- **Model** the action of a glandular stomach.
- **Model** the action of a muscular stomach.
- **Compare** the stomach models.

MATERIALS

- safety goggles
- disposable gloves
- graduated cylinder, 50 mL
- forceps
- sand paper
- aquarium gravel
- lab apron
- glass beakers, small (2)
- vinegar
- dustless board chalk (broken into ¼ sticks)
- felt (2 dark-colored squares)
- hand lens

Procedure

MODEL THE ACTION OF THE GLANDULAR STOMACH

1. Put on safety goggles, a lab apron, and gloves.

2. **CAUTION: Use glass beakers with care.** Add about 50 mL of vinegar to two small beakers.
   Note: Vinegar is a weak acid. Exercise caution when using vinegar.

3. Use your forceps to carefully add a piece of chalk to the beaker. Observe and record any changes in the appearance of the chalk.

   _____
   _____
   _____

4. Use a piece of sand paper to grate the dustless chalk into small particles.
   Note: Do not inhale these particles.

Name _____ Class _____ Date _____

Bird Digestion *continued*

5. Add the particles to the second beaker. Record your observations.

_____

_____

_____

_____

MODEL THE ACTION OF THE GLANDULAR STOMACH

6. Place a piece of chalk in the center of a clean, square of felt. Close up the fabric around the chalk.

7. Squeeze and roll the chalk against a hard surface for 20 seconds. Open the fabric and examine the appearance of the chalk and felt. Use a hand lens to examine any particles that may have been produced.

8. Place a fresh piece of chalk in the center of the second felt square. Add a teaspoon of aquarium gravel and close up the fabric.

9. Squeeze and roll the chalk against a hard surface for 20 seconds.

10. Open the fabric and examine the appearance of the chalk, gravel, and felt. Use a hand lens to examine any particles that may have been produced.

11. Clean up your lab materials according to your teacher's instructions. Wash your hands before leaving the lab.

Analyze and Conclude

1. **Identifying Variables** In the first part of this activity, what did the chalk and vinegar represent?

_____

_____

2. **Evaluating Results** What happened when the chalk was added to the vinegar solution?

_____

_____

3. **Analyzing Methods** How did grating the chalk into smaller particles affect the reaction?

_____

_____

_____

Name _____ Class _____ Date _____

Bird Digestion *continued*

4. **Identifying Variables** In the second part of this activity, what did the felt, gravel, and chalk represent?

   _____
   _____

5. **Using Evidence to Develop Explanations** How did adding gravel to the chalk affect its breakdown?

   _____
   _____

6. **Inferring Conclusions** In some birds, the digesting food is often shifted back and forth between both stomachs. What advantage does this offer?

   _____
   _____
   _____

7. **Forming Reasoned Opinions** Gizzard stones that become rounded and smooth often pass out of the stomach. The bird must then ingest more stones to replenish this load. What advantage might this have?

   _____
   _____
   _____

Extensions

8. **Further Inquiry** Write a new question about bird digestion that could be explored with another investigation.

   _____
   _____
   _____
   _____

Bird Digestion *continued*

9. **Designing Models** Choose another bird organ system, such as a bird's respiratory system or reproductive system. Research how the system functions and then describe how you could build a model of that system. Include a list of materials that you would need in your plan.

Name _____ Class _____ Date _____

Skills Practice

DATASHEET B FOR IN-TEXT LAB

Mammalian Characteristics

OBJECTIVES

- **Examine** distinguishing characteristics of mammals.
- **Infer** the functions of mammalian structures.

MATERIALS

- disposable gloves
- prepared slide of mammalian skin
- mirror
- hand lens or stereomicroscope
- compound microscope
- specimens or pictures of vertebrate skulls (some mammalian and some nonmammalian)

Procedure

1. Use the data table below or make your own.

Mammalian Teeth

Mammal	Incisors	Canines	Premolars	Molars

Name _____ Class _____ Date _____
Mammalian Characteristics *continued*

2. Write a question that you would like to explore about the characteristics of mammals based on the objectives for this lab.

   _____

   _____

3. Use a hand lens to look at several areas of your skin, including areas that appear to be hairless. Record your observations.

   _____

   _____

   _____

   _____

4. Look at a prepared slide of mammalian skin under low power of a compound microscope. Notice the glands in the skin. Look for the oil (sebaceous) glands and the sweat glands. Draw and label an example of each type of gland.

Name _____ Class _____ Date _____

Mammalian Characteristics *continued*

5. **CAUTION: Wash your hands thoroughly with soap and water.** Use a mirror to look in your mouth. Identify the four kinds of mammalian teeth that are in your mouth.

   _____

   _____

6. Count each kind of tooth on one side of your lower jaw. Multiply the number of each kind of tooth by 4, and record your results in the appropriate columns of your data table. Before continuing, wash your hands again.

7. Look at the skulls of several mammals. Identify the kinds of teeth in each skull. For each skull, find the number of each kind of tooth. Record your results in your data table.

8. Look at the skulls of several nonmammalian vertebrates, and compare nonmammalian teeth to mammalian teeth.

   _____

   _____

   _____

   _____

9. Compare the jaws of mammalian skulls to those of nonmammalian vertebrates. As you look at each skull, notice the structure of the lower jawbone and the way in which the upper jawbone and the lower jawbone connect.

10. Clean up your lab materials according to your teacher's instructions. Wash your hands before leaving the lab.

Analyze and Conclude

1. **Summarizing Information** List the characteristics that distinguish mammals from other vertebrates.

   _____

   _____

   _____

   _____

   _____

   _____

Mammalian Characteristics *continued*

2. **Inferring Relationships** What role, if any, might hair or fur play in enabling mammals to be endotherms?

   _____
   _____
   _____
   _____

3. **Scientific Methods Forming Hypothesis** What roles other than the role identified in item 2 above do you think hair might play in mammals?

   _____
   _____
   _____
   _____

4. **Recognizing Patterns** Where are the oil (sebaceous) glands located in the skin of mammals?

   _____

5. **Comparing Structures** How does the mammalian jaw differ from nonmammalian jaws?

   _____
   _____

6. **Scientific Methods Inferring Conclusions** Given the shape of your teeth, would you classify humans as carnivores (meat eaters), herbivores (plant eaters), or omnivores (meat and plant eaters)? Explain.

   _____
   _____
   _____
   _____

Name _____ Class _____ Date _____

Mammalian Characteristics *continued*

7. **Scientific Methods Evaluating Conclusions** Justify the following conclusion: The kinds and shapes of a mammal's teeth can be used to determine the mammal's diet.

_____

_____

_____

Extensions

8. **Further Inquiry** Write a new question about the characteristics of mammals that could be explored in a new investigation.

_____

_____

Name _____ Class _____ Date _____

Inquiry Lab

DATASHEET B FOR IN-TEXT LAB

Territorial Behavior

OBJECTIVES

- **Recognize** that territorial behavior is a type of social behavior.
- **Observe** how male crickets behave in close proximity to a defendable resource, such as food.
- **Form hypotheses** about the function of male territorial behavior in crickets.

MATERIALS

- disposable gloves
- colored paint, nontoxic, washable
- aquarium
- cardboard, 5 cm square
- potato, piece
- crickets (5 male and 5 female)
- paint brushes, several, small
- cardboard tube
- apple, slice
- watch, with second hand

Preparation

1. **Scientific Methods State the Problem** Under what circumstances do male crickets chirp most often?

   _____

   _____

2. **Scientific Methods Form a Hypothesis** Form a testable hypothesis that explains how different situations and cues trigger chirping (territorial behavior) in a male cricket. Record your hypothesis.

   _____

   _____

Procedure

OBSERVE TERRITORIAL BEHAVIOR

1. Put on gloves.

2.   **CAUTION: Crickets are animals and should be handled with care.** With colored paint, mark the backs of 5 male crickets. Use a different color for each cricket. Place the crickets in an aquarium.

3. Place 5 unmarked female crickets in the aquarium.

Territorial Behavior *continued*

4. Make two shelters. Construct the first shelter by turning the cardboard tube on its side. Construct the second shelter by folding the cardboard square in half to form a tent-like structure.
5. Place the shelters in different spots in the aquarium.
6. Place a slice of apple and a piece of potato in the aquarium.
7. Observe the crickets for 10 minutes. Look for territorial behaviors among the males, such as chirping, stroking others with antennae, and pushing others away.

DESIGN AN EXPERIMENT

8. Design an experiment that tests your hypothesis and that uses the materials listed for this lab. Predict what will happen during your experiment if your hypothesis is supported.

_____

_____

_____

_____

9. Write a procedure for your experiment. Identify the variables that you will control, the experimental variables, and the responding variables.

_____

_____

_____

_____

Construct any tables you will need to record your data. Make a list of all safety precautions that you will take. Have your teacher approve your procedure before you begin.

CONDUCT YOUR EXPERIMENT

10. Put on gloves. Carry out your experiment. Record your observations in the data table.
11. 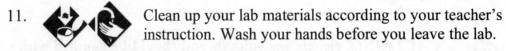 Clean up your lab materials according to your teacher's instruction. Wash your hands before you leave the lab.

Territorial Behavior *continued*

Analyze and Conclude

1. **Scientific Methods** **Analyzing Data** Were any crickets more aggressive than the others? Give evidence to support your answer.

2. **Describing Results** What were the circumstances in which most aggressive behavior occurred? Was your prediction correct?

3. **Scientific Methods** **Identifying Variables** What did the shelter and food represent to the male crickets?

4. **Scientific Methods** **Forming Hypotheses** For each aggressive behavior observed, form a hypothesis that explains the behavior's function.

Extensions

5. **Insect Behavior** Use the Internet to research the types of territorial behavior in other insects. Are the behaviors observed in the crickets typical for other insects?

Name _____ Class _____ Date _____

Inquiry Lab   DATASHEET B FOR IN-TEXT LAB

Analysis of Muscle Fatigue

OBJECTIVES
- **Relate** muscle energy use and fatigue to the work muscles do.
- **Observe** the effects of fatigue on a muscle.

MATERIALS
- hand grips, spring
- graph paper
- watch with second hand

Preparation

1. **Scientific Methods** **State the Problem** How do dominant and non-dominant hands compare in the amount of work they can do?

   _____

   _____

2. **Scientific Methods** **Form a Hypothesis** Form a testable hypothesis that predicts the amount of work that can be done by each hand.

   _____

   _____

   _____

   _____

Procedure

1. Design an experiment that tests your hypothesis and that uses the materials listed for this lab.

   _____

   _____

   _____

   _____

Holt Biology — Systems

Analysis of Muscle Fatigue *continued*

2. Write a procedure for your experiment. Identify the variables that you will control, the experimental variables, and the responding variables. You may use the table below to record your results, or make one of your own.

_____

_____

_____

_____

_____

_____

Trial	Squeezes

3. Make a list of all of the safety precautions that you will take. Have your teacher approve your procedure before you begin.

_____

_____

Analyze and Conclude

1. **Summarizing Results** Using your graph paper, make a graph of your data. Use the *x*-axis for the number of trials. Use the *y*-axis for the number of muscle contractions (squeezes).

Name _____ Class _____ Date _____
Analysis of Muscle Fatigue *continued*

2. **Scientific Methods Analyzing Results** Explain the differences in the amount of work done by the muscles during the three trials.

_____
_____
_____
_____

3. **Scientific Methods Drawing Conclusions** What is the relationship between the work muscles can do and fatigue?

_____
_____

4. **Comparing Functions** Compare the work done by muscles in your hand to the work done by your heart muscle.

_____
_____
_____
_____
_____

Name _____ Class _____ Date _____

Skills Practice

DATASHEET B FOR IN-TEXT LAB

Lung Capacity

OBJECTIVES

- **Measure** the components that make up your lung capacity, such as your tidal volume, vital capacity, and expiratory reserve volume.
- **Determine** your inspiratory reserve volume and your lung capacity.
- **Predict** how exercise will affect tidal volume, vital capacity, and lung capacity.

MATERIALS

- spirometer
- spirometer mouthpiece

Procedure

MEASURE AIR VOLUME

1. Copy the data table shown into your lab notebook.
2. **CAUTION: Place a clean mouthpiece in the head of a spirometer.** To measure your tidal volume, first inhale a normal breath. Then exhale a normal breath into the spirometer through the mouthpiece. Record the volume of air exhaled in your data table. Many diseases are spread by body fluids, such as saliva. Note: Do not share a spirometer mouthpiece with anyone.

Spirometer Readings

Tidal volume (TV)	
Expiratory reserve volume (ERV)	
Inspiratory reserve volume (IRV)	
Vital capacity (VC)	
Estimated residual volume (RV)	
Estimated lung capacity (LC)	

Name _____ Class _____ Date _____

Lung Capacity *continued*

3. To measure your expiratory reserve volume, inhale a normal breath and then exhale normally. Then forcefully exhale as much air as possible into the spirometer. Record this volume.

4. To measure your vital capacity, first inhale as much as you can. Then forcefully exhale as much air as you can into the spirometer. Record this volume.

CALCULATE LUNG CAPACITY

5. The data table shown on the top of page 883 in your textbook contains average values for residual volume and lung capacity for young adults. Inspiratory reserve volume (IRV) can be calculated by subtracting tidal volume (TV) and expiratory reserve volume (ERV) from vital capacity (VC). The equation for this calculation is as follows: IRV = VC − TV − ERV. Use this data in your data table and the equation above to calculate your estimated inspiratory reserve volume.

   Estimated inspiratory reserve volume = _____

6. Lung capacity (LC) can be calculated by adding residual volume (RV) to vital capacity (VC). The equation for this calculation is as follows: LC = VC + RV. Use the data in your data table above to calculate your estimated lung capacity.

   Estimated lung capacity = _____

   Compare your estimated lung capacity to the value for lung capacity on page 883.

7. Clean up your lab materials according to your teacher's instructions. Wash your hands thoroughly before leaving the lab.

Analyze and Conclude

1. **Interpreting Data** How does your expiratory reserve volume compare with your inspiratory reserve volume?

   _____

   _____

2. **Interpreting Tables** How do the average values for residual volume and lung capacity for young adult females compare with the average values for young adult males?

   _____

   _____

   _____

Name _____ Class _____ Date _____

Lung Capacity *continued*

3. **Analyzing Data** How did your tidal volume compare with the volumes of others?

   _____
   _____

4. **Recognizing Relationships** Why was the value that you found for your lung capacity an estimated value?

   _____
   _____

5. **Scientific Methods Analyzing Methods** Why did you measure inspiratory reserve volume indirectly?

   _____
   _____

6. **Scientific Methods Inferring Conclusions** Why do males and athletes have greater vital capacities than females do?

   _____
   _____

7. **Scientific Methods Justifying Conclusions** Use the data from your class to justify the conclusion that exercise increases lung capacity.

   _____
   _____
   _____
   _____

Name _____ Class _____ Date _____

Lung Capacity *continued*

Extensions

8. **Respiratory Disease** Spirometry is the use of a spirometer to study respiratory function. Nurses and respiratory therapists use spirometers to evaluate patients who have respiratory diseases. Do research to discover how spirometry is used to distinguish different respiratory diseases. Make a chart showing what you discover.

9. **Further Inquiry** Write a new question that could be explored with another investigation.

_____

_____

_____

Name _____ Class _____ Date _____

Inquiry

DATASHEET B FOR IN-TEXT LAB

Lactose Digestion

OBJECTIVES

• **Describe** the relationship between enzymes and the digestion of food molecules.

• **Evaluate** the ability of the lactase enzyme to promote lactose digestion.

• **Infer** the presence of lactose in milk and foods that contain milk.

MATERIALS

- safety goggles
- lab apron
- toothpicks
- droppers
- glucose solution
- disposable gloves
- lactase enzyme
- spot plates
- whole milk
- glucose test strips

Preparation

1. **Scientific Methods State the Problem** How does the lactase enzyme aid in the digestion of lactose?

2. **Scientific Methods Form a Hypothesis** Form a testable hypothesis that explains how the lactase enzyme affects milk.

_____

_____

Holt Biology   Digestive and Excretory Systems

Name _____ Class _____ Date _____

Lactose Digestion *continued*

Procedure

LEARN ABOUT LAB MATERIALS

1. Glucose test strips can be used to indicate the presence or absence of glucose in a solution. To use the test strip, touch the test pad to the solution being tested. After the suggested amount of time, compare the test strip to the color guide.

2. Do some research to find out how the lactase enzyme works. Discuss your findings with your lab group. Below, write a summary of what you have learned.

_____
_____
_____
_____

DESIGN AN EXPERIMENT

3. Design an experiment that tests your hypothesis and that uses the materials listed for this lab. Predict what will happen during your experiment if your hypothesis is supported.

_____
_____
_____

4. Write a procedure for your experiment. Identify the variables that you will control, the experimental variables, and the responding variables. Construct any tables you will need to record your data. Make a list of all the safety precautions that you will take. Have your teacher approve your procedure before you begin the experiment.

_____
_____
_____
_____

CONDUCT YOUR EXPERIMENT

5.  Put on a lab apron, safety goggles, and gloves.

Name _____ Class _____ Date _____

Lactose Digestion *continued*

6. **CAUTION: Handle glass slides with care. Do not touch or taste any chemicals.** Carry out your experiment. Observe the test tubes after 24 hours. Record your data in a data table.

7. Clean up your lab materials according to your teacher's instructions. Wash your hands before leaving the lab.

Analyze and Conclude

1. **Summarizing Information** Did the lactase enzyme have an effect on the milk?

   _____
   _____

2. **Recognizing Relationships** What is the relationship between lactose and lactase?

   _____
   _____
   _____

3. **Scientific Methods Interpreting Data** What role did the glucose solution play in your experiment?

   _____
   _____
   _____

4. **Drawing Conclusions** What does the lactase enzyme do to milk?

   _____
   _____

5. **Scientific Methods Using Evidence to Make Explanations** How do your results justify your conclusions?

   _____
   _____
   _____

Name _____ Class _____ Date _____

Lactose Digestion *continued*

6. **Scientific Methods Controlling Variables** Why should you test the lactase enzyme with glucose test strips?

   _____

   _____

7. **Further Inquiry** Write a new question about enzymes and digestion that could be explored with another investigation.

   _____

   _____

Extensions

8. **Recognizing Patterns** What are some other food-treatment products that contain digestive enzymes?

   _____

   _____

   _____

9. **Inferring Conclusions** Why does the improper breakdown of certain food molecules cause symptoms such as stomach pain, gas, and diarrhea?

   _____

   _____

   _____

Name_____ Class_____ Date_____

Skills Practice Lab　　　　　　　　　　　DATASHEET B FOR IN-TEXT LAB

Disease Transmission Model

OBJECTIVES
- **Model** disease transmission.
- **Organize** and **analyze** data.

MATERIALS
- test tube, large
- dropper bottle of iodophenol indicator
- dropper bottle of unknown solution

Procedure
SIMULATE DISEASE TRANSMISSION
1. Use the data tables below as the lab progresses.

Disease Transmission

Round number	Partner's name

Disease Source

Name of infected person	Names of infected person's partners		
	Round 1	Round 2	Round 3

Disease Transmission Model *continued*

2. Put on disposable gloves, a lab apron, and safety goggles.

3. **CAUTION: Do not touch or taste any chemicals. Exercise caution when working with glassware such as a test tube.**
   You will be given a dropper bottle of an unknown solution. When your teacher says to begin, transfer three droppersful of your solution to a clean test tube.

4. Select a partner for Round 1. Record the name of this partner in your Disease Transmission table.

5. Together, you and your partner have two test tubes. Pour the contents of one of the test tubes into the other test tube. Then, pour half of the solution back into the first test tube. You and your partner now share any pathogens either of you might have.

6. On your teacher's signal, select a new partner for Round 2. Record the partner's name in your Disease Transmission table. Repeat step 5.

7. On your teacher's signal, select a new partner for Round 3. Record the partner's name in your Disease Transmission table. Repeat step 5.

8. Add one dropperful of indophenol indicator to your test tube. "Infected" solutions will be colorless or will turn light pink. "Uninfected" solutions will turn blue. Record the results of your test.

TRACE THE DISEASE SOURCE

9. If you are infected, write your name and the name of your partner in each round on the board or on an overhead projector. Mark your infected partners. Record all of the data for your class in your Disease Source table.

10. To trace the source of the infection, cross out the names of the uninfected partners in Round 1. There should be only two names left. One is the name of the original disease carrier. To find the original disease carrier, place a sample from his or her dropper bottle in a clean test tube, and test it with indophenol indicator.

Name _____ Class _____ Date _____

Disease Transmission Model *continued*

11. To show the disease transmission route, use the diagram below. Show the original disease carrier and the people each disease carrier infected.

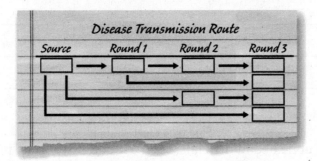

12. Clean up your lab materials according to your teacher's instructions. Wash your hands thoroughly before leaving the lab.

Analyze and Conclude

1. **Interpreting Data** After Round 3, how many people were infected? Express this number as a percentage of your class.

   _____

2. **Relating Concepts** What do you think the clear fluids each student started with represent? Explain your response.

   _____
   _____
   _____
   _____

3. **Drawing Conclusions** Can someone who does not show any symptoms of a disease transmit that disease? Explain.

   _____
   _____
   _____
   _____
   _____
   _____

Name _____ Class _____ Date _____

Disease Transmission Model *continued*

4. **Further Inquiry** Write a new question about disease transmission that could be explored with another investigation.

_____
_____
_____

Extensions

5. **On the Job** Public health officials, such as food inspectors, research and work to stop the spread of diseases in human populations. Do research to find out how public health officials trace the origin of communicable diseases.

_____
_____
_____
_____
_____
_____
_____
_____

Name _____ Class _____ Date _____

Inquiry Lab

DATASHEET B FOR IN-TEXT LAB

Reaction Times

OBJECTIVES
- **Determine** human reaction times.
- **Design** an experiment that measures changes in reaction times.

MATERIALS
- meterstick

Preparation

1. **Scientific Methods State the Problem** What factors might influence how fast your reaction time is in response to a task?
2. **Scientific Methods Form a Hypothesis** Form a testable hypothesis that explains how a factor might influence how fast reaction time is in response to a task. Record your hypothesis.

_____

_____

_____

Procedure

CALCULATE REACTION TIMES

1. Use the data table on the next page to record reactions.
2. Sit in a chair, and have a partner stand facing you while holding a meterstick in a vertical position.
3. Hold your thumb about 3 cm from your fingers near the bottom end of the stick. The meterstick should be positioned to fall between your thumb and fingers.
4. Tell your partner to let go of the meterstick without warning you.
5. When your partner releases the meterstick, catch the stick by pressing your thumb and fingers together. Your partner should catch the top of the stick if it begins to tip over.

Original content Copyright © by Holt, Rinehart and Winston. Additions and changes to the original content are the responsibility of the instructor.

Holt Biology   Nervous System

Name _____ Class _____ Date _____

Reaction Times *continued*

6. Record the number of centimeters that the stick dropped before you caught it. The distance that the meterstick fell can be used to evaluate your reaction time.

Reactions

Hand: trial number	Subject 1 reaction distance (centimeters, cm)	Subject 2 reaction distance (centimeters, cm)
Left: 1		
Left: 2		
Left: 3		
Left: average		
Right: 1		
Right: 2		
Right: 3		
Right: average		

7. Repeat the procedure several times, and calculate the average number of centimeters that the meterstick dropped during each trial.

8. Try this procedure with your other hand. Close your eyes and have your partner say "now" when the stick is released.

9. Exchange places with your partner, and repeat the procedure.

DESIGN AN EXPERIMENT

10. Design an experiment that tests your hypothesis and that uses the materials listed for the lab. Predict what will happen during your experiment if your hypothesis is supported.

_____
_____
_____
_____
_____

Name _____ Class _____ Date _____
Reaction Times *continued*

11. Write a procedure for your experiment. Identify the variables that you will control, the experimental variables, and the responding variables. Construct any tables you will need to record your data. Make a list of all the safety precautions you will take. Have your teacher approve your procedure before you begin.

_____
_____
_____
_____
_____
_____

CONDUCT YOUR EXPERIMENT

12. Set up your group's experiment, and collect data.
13. Clean up your lab materials according to your teacher's instructions. Wash your hands before leaving the lab.

Analyze and Conclude

1. **Summarizing Results** What was your shortest reaction time?

_____
_____

2. **Scientific Methods Analyzing Data** How does your reaction when you use your dominant hand compare with your reaction when you use your other hand?

_____
_____

3. **Scientific Methods Drawing Conclusions** Why may each hand have a different reaction distance? Why may each person have a different reaction distance?

_____
_____
_____
_____

Original content Copyright © by Holt, Rinehart and Winston. Additions and changes to the original content are the responsibility of the instructor.

Holt Biology — Nervous System

Name _____ Class _____ Date _____

Reaction Times *continued*

4. **Predicting Patterns** Compile the data gathered by each lab group in your class. Can you identify any trends in the data?

   _____
   _____
   _____
   _____

5. **Scientific Methods Analyzing Results** How did the data differ in the experiment that you designed? Explain whether or not your hypothesis was supported.

   _____
   _____
   _____
   _____

Extensions

6. **Predicting Outcomes** Do research in the library or media center to determine how athletes can improve their reaction times. Are these methods in common practice at your school?

   _____
   _____
   _____
   _____
   _____

7. **Identifying Concepts** What factors can influence how fast a person reacts to a stimulus?

   _____
   _____
   _____
   _____

Name _____ Class _____ Date _____
Reaction Times *continued*

8. **Further Inquiry** Write a new question about reaction times that could be explored in another investigation.

_____

_____

_____

_____

Name _____ Class _____ Date _____

Inquiry   DATASHEET B FOR IN-TEXT LAB

Epinephrine and Heart Rate

OBJECTIVES

- **Measure** the heart rate of daphnia.
- **Observe** the effect of the hormone epinephrine on heart rate in daphnia.
- **Determine** the threshold concentration for the effects of epinephrine on daphnia.

MATERIALS

- medicine dropper
- daphnia
- daphnia culture water
- depression slide
- petroleum jelly
- epinephrine solutions, 0.001%, 0.0001%, 0.00001%, and 0.000001%
- compound microscope
- watch or clock with second hand
- paper towels
- beaker, 100 mL
- graduated cylinder, 10 mL
- coverslip

Preparation

1. **Scientific Methods State the Problem** What is the threshold concentration of epinephrine that affects the heart rate of daphnia?

   _____

2. **Scientific Methods Form a Hypothesis** Form a testable hypothesis that explains how epinephrine might affect the heart rate of daphnia. Record your hypothesis.

   _____
   _____

Procedure

OBSERVE HEART RATE IN DAPHNIA

1.  Put on safety goggles, gloves, and a lab apron.

Name _____ Class _____ Date _____

Epinephrine and Heart Rate *continued*

2. **CAUTION: Do not touch your face while handling microorganisms.** Use a clean medicine dropper to transfer one daphnia to the well of a clean depression slide. Place a dab of petroleum jelly in the well. Add a coverslip. Observe the daphnia with a compound microscope under low power.

3. Count the daphnia's heartbeats for 10 s. Divide this number by 10 to find the HR in beats per second. Record your observations in the data table as Trial 1. Turn off the microscope light, and wait 20 s. Repeat this step two times, and record your observations as Trials 2 and 3.

Daphnia Heart Rate

Solution	HR (beats per second) Trial 1 (A)	HR (beats per second) Trial 2 (B)	HR (beats per second) Trial 3 (C)	Average HR (beats per second) [(A+B+C)/3]	Average HR (beats per minute)

4. Calculate the average HR in beats per second. Then calculate the average heart rate in beats per minute using the following formula:

   Ave HR (beats per minute) = Ave HR (beats per second) × 60 s/min.

DESIGN AN EXPERIMENT

5. Design an experiment that tests your hypothesis and that uses the materials listed for this lab. Predict what will happen during your experiment if your hypothesis is supported.

_____

_____

_____

Name _____ Class _____ Date _____
Epinephrine and Heart Rate *continued*

6. **CAUTION: Handle animals carefully and with respect.** Write a procedure for your experiment. Identify the controlled variables, the experimental variables, and the responding variables. Construct any tables that you will need to record your data. Make a list of all of the safety precautions that you will take. Have your teacher approve your procedure and safety precautions before you begin.

_____

_____

_____

_____

CONDUCT YOUR EXPERIMENT

7. **CAUTION: Glassware such as coverslips and slides are fragile. Notify your teacher of broken glass or cuts.** To add a solution to the prepared slide, first place a drop of the solution at the edge of the coverslip. Then, place a piece of paper towel along the opposite edge to draw the solution under the coverslip. Wait 1 min for the solution to take effect.

8. **CAUTION: Epinephrine is toxic and is absorbed through the skin. Wear gloves at all times during this experiment. Notify your teacher of any spills.** Perform your experiment. Record your observations in a data table.

9. Clean up your lab materials according to your teacher's instructions. Wash your hands before leaving the lab.

Analyze and Conclude

1. **Summarizing Results** Make a graph of your data. Plot "Epinephrine concentration (%)" on the *x*-axis. Plot "Average heart rate (beats per minute)" on the *y*-axis.

2. **Scientific Methods Analyzing Data** Which solutions affected the heart rate of daphnia?

_____

_____

Name _____ Class _____ Date _____

Epinephrine and Heart Rate *continued*

3. **Scientific Methods Interpreting Data** What was the threshold concentration of epinephrine?

   _____
   _____
   _____

4. **Making Predictions** Based on the information you have and based on your data, predict how epinephrine concentration would affect heart rates in humans.

   _____
   _____
   _____
   _____
   _____

Extensions

5. **Inferring Relationships** Research anaphylactic shock. Explain why epinephrine is used to treat anaphylactic shock.

   _____
   _____
   _____
   _____
   _____
   _____
   _____

6. **Further Inquiry** Write a new question about hormones that could be explored with another investigation.

   _____
   _____

Name _____ Class _____ Date _____

Skills Practice

Sonography

DATASHEET B FOR IN-TEXT LAB

OBJECTIVES

- **Model** how a fetus can be seen using sonography.
- **Analyze** how waves move through different materials.

MATERIALS

- lab apron
- metric ruler
- corn syrup
- large spoon
- cookie tray
- water, tap
- timer
- various objects such as a wooden block, a domino, and paper clips

Procedure

MEASURE AIR VOLUME

1. Put on a lab apron.

2. Place a cookie tray on a flat, level surface. Carefully fill it with water to a depth of about 0.5 cm.

3. Position a spoon at one end of the tray. Keeping the spoon level over the water's surface, gently tap down once on the water. Be sure that your fingertips don't extend into the water. What do you observe?

    _____

    _____

    How long does it take for the return wave to reach the spoon? _____

    Continue tapping to form a regular rhythm.

4. Place a domino or other target with a flat edge midway in the tray.

Original content Copyright © by Holt, Rinehart and Winston. Additions and changes to the original content are the responsibility of the instructor.

Holt Biology · 162 · Reproduction and Development

Name _____ Class _____ Date _____

Sonography *continued*

5. Tap out a regular rhythm. What do you observe?

   _____

   _____

   Tap down once. How long does it take for the return wave to reach the spoon?

   _____

6. Move the target closer to the spoon. How does this affect the time needed for the return signal?

   _____

   _____

   Place the target further away from the spoon. What happens now?

   _____

   _____

7. Use a variety of targets with different shapes including curved and angular edges. Note how these targets affect the return wave.

   _____

   _____

8. Place several drops of corn syrup in the center of the tray. Observe how this more dense liquid affects the waves that pass through it.

   _____

   _____

9. Clean up your lab materials according to your teacher's instructions. Wash your hands thoroughly before leaving the lab.

Analyze and Conclude

1. **Summarizing Results** How did the spoon's impact affect the water's surface?

   _____

   _____

Name _____ Class _____ Date _____
Sonography *continued*

2. **Describing Observations** What happened when this wave struck a hard surface?

   _____
   _____
   _____

3. **Scientific Methods Drawing Conclusions** How did the distance to the target affect the wave's return?

   _____
   _____
   _____

4. **Evaluating Models** How does this activity model the use of sonography in prenatal observation?

   _____
   _____
   _____
   _____

5. **Scientific Methods Analyzing Results** How did traveling through the corn syrup affect the wave? How might this be applied to sonography?

   _____
   _____
   _____
   _____

Name _____ Class _____ Date _____

Skills Practice   DATASHEET B FOR IN-TEXT LAB

The Counterfeit Drug

OBJECTIVES

- **Practice** using the technique of paper chromatography to separate pigments.
- **Determine** whether a dye contains one or two pigments.

MATERIALS

- scissors
- tape
- beakers, small (2)
- toothpicks (2)
- water, distilled
- chromatography paper (2 strips)
- pencils (2)
- paper clips, small (2)
- Pigment solutions to be tested

Procedure

1.  Put on safety goggles, gloves, and a lab apron.

2. **CAUTION: Do not use chipped or cracked glassware. Notify your teacher immediately if a piece of glassware breaks. Use extreme care when handling scissors.** Cut two lengths of chromatography paper equal to the depth of the beakers that you are using. (Hint: Check for length by taping the top of the strip to a pencil. Lower the paper into the beaker. Let the pencil rest across the top of the beaker.)

3. Once the paper is cut to the correct length, remove it from the beaker. Leave the pencil attached to the top of the paper.

4. Attach a paper clip to the bottom of each strip of paper in order to keep the paper hanging straight down while it is in the beaker.

5. **CAUTION: Dyes will stain.** Do not get any dye on your skin or your clothing. By using a toothpick, dab some of the unknown dye onto one strip of chromatography paper. Add the dye about 2 cm above where the water level will be in the bottom of the beaker.

Name _____ Class _____ Date _____

The Counterfeit Drug *continued*

6. Allow the dye to dry for a few minutes. Repeat the dabbing process several times (at the same place on the paper each time) to build up a concentration of pigment that will yield good results.

7. Repeat steps 5 and 6 using the control dye, which was supplied by the legitimate manufacturer of the pills.

8. Pour distilled water into each beaker to a depth of about 2 cm.

9. Make sure that the dye is above the water level and does not come into contact with the water. Refer to the photo on p. 1037 of your textbook, which shows the correct distance between the dye spot and the water level.

10. Set up the beakers, paper, pencils, and water as shown in the photo. Allow the paper to absorb water so that the water moves up the paper by capillary action and separates the pigments in the dye. This process may take several minutes.

11. When the water migrates to within a few millimeters from the top of the paper, remove the pencils and paper from each beaker.

12. Empty the water from the beaker, place the pencils back across the top of each beaker, and allow your chromatograms to dry overnight.

13. Clean up your lab materials according to your teacher's instruction. Wash your hands before you leave the lab.

Analyze and Conclude

1. **Drawing Conclusions** Did the unknown sample come from the genuine drug or the counterfeit drug? How can you tell?

   _____

   _____

2. **Explaining Events** Why do the different pigments that make up a dye separate on the chromatography paper?

   _____

   _____

   _____

   _____

3. **Critical Thinking Analyzing Methods** Why are tests in cases like this one carried out in such a way that the analysts do not know what products they are testing?

   _____

   _____

   _____

The Counterfeit Drug *continued*

Extensions

4. **Forensic Careers** Research and report on careers for forensic scientists in the FDA or DEA.

_____

_____

_____

_____

_____

_____

_____

_____